Name _____ Class _____

Skills Worksheet
Directed Reading

Section: Meiosis

Read each question, and write your answer in the space provided.

1. What is meiosis?

   _____
   _____
   _____
   _____

2. Explain the difference between meiosis I and meiosis II.

   _____
   _____
   _____
   _____

3. List the stages of meiosis in the order that they occur.

   _____
   _____
   _____
   _____

4. What is crossing-over?

   _____
   _____
   _____
   _____

Copyright © by Holt, Rinehart and Winston. All rights reserved.

Holt Biology   1   Meiosis and Sexual Reproduction

Directed Reading continued

In the space provided, write the name of the stage of meiosis that is being described.

_____ 5. The centromeres divide, and the chromatids move to opposite poles of the cell.

_____ 6. The homologous chromosomes separate. The chromosomes of each pair are pulled to opposite poles of the cell by the spindle fibers. The chromatids do not separate at their centromeres.

_____ 7. The chromosomes condense, and the nuclear envelope breaks down. Homologous chromosomes pair all along their length and then cross over.

_____ 8. After one division of the nucleus, a new spindle forms around each group of chromosomes.

_____ 9. Individual chromosomes line up along the equator, attached at their centromeres to spindle fibers.

_____ 10. A nuclear envelope forms around each set of chromosomes. Two cells undergo cytokinesis, forming haploid offspring cells.

_____ 11. Individual chromosomes gather at each of the two poles. In most organisms, the cytoplasm divides, forming two new cells.

_____ 12. The pairs of homologous chromosomes are moved by the spindle to the equator of the cell. The homologous chromosomes, each made up of two chromatids, remain together.

Name _____ Class _____ Date _____

Directed Reading continued

Read each question, and write your answer in the space provided.

13. What is crossing-over? During which phase of meiosis does crossing-over occur?

    _____

    _____

14. What is independent assortment? During which phase(s) of meiosis does independent assortment occur?

    _____

    _____

15. What are spermatogenesis and oogenesis?

    _____

    _____

16. What is the difference between undifferentiated sperm cells and sperm?

    _____

    _____

17. Why does meiosis produce four sperm cells but only one ovum?

    _____

    _____

    _____

Skills Worksheet

Directed Reading

Section: Sexual Reproduction

In the space provided, explain how the terms in each pair differ in meaning.

1. asexual reproduction, sexual reproduction

   _____

   _____

   _____

2. clone, asexual reproduction

   _____

   _____

   _____

3. fission, budding

   _____

   _____

   _____

4. budding, fragmentation

   _____

   _____

   _____

Read each question, and write your answer in the space provided.

5. What are the advantages and disadvantages of asexual reproduction?

   _____

   _____

   _____

6. What are the advantages and disadvantages of sexual reproduction?

   _____

   _____

   _____

Copyright © by Holt, Rinehart and Winston. All rights reserved.

Holt Biology   Meiosis and Sexual Reproduction

Directed Reading *continued*

7. List the three types of life cycles that a eukaryote that undergoes sexual reproduction can have. Give an example of an organism that undergoes each type of life cycle.

_____

_____

_____

8. How can you tell which type of sexual life cycle an organism undergoes?

_____

_____

_____

9. List the main steps in the diploid life cycle. Begin with meiosis.

_____

_____

_____

Name _____ Class _____ Date _____

Skills Worksheet

Active Reading

Section: Meiosis

Read the passage below. Then answer the questions that follow.

Meiosis is a form of cell division that halves the number of chromosomes when forming specialized reproductive cells, such as gametes or spores. Meiosis involves two divisions of the nucleus—meiosis I and meiosis II.

The stages of meiosis I are as follows:

Prophase I: The chromosomes condense, and the nuclear envelope breaks down. Homologous chromosomes pair along their length and then cross over.

Metaphase I: The pairs of homologous chromosomes are moved by the spindle to the equator of the cell. The homologous chromosomes, each made up of two chromatids, remain together.

Anaphase I: The homologous chromosomes separate. As in mitosis, the chromosomes of each pair are pulled to opposite poles of the cell by the spindle fibers. But in meiosis, the chromatids do not separate at their centromeres.

Telophase I: Individual chromosomes gather at each of the poles. In most organisms, the cytoplasm divides, forming two new cells.

SKILL: READING EFFECTIVELY

Match each statement with the stage of meiosis I it describes by writing in the spaces provided, *PI* to represent Prophase I, *MI* to represent Metaphase I, *AI* to represent Anaphase I, or *TI* to represent Telophase I.

_____ 1. cytoplasm divides

_____ 2. nuclear envelope breaks down

_____ 3. homologous chromosomes separate

_____ 4. spindle moves homologous chromosomes to the cell's equator

_____ 5. crossing-over occurs

_____ 6. two new cells form

_____ 7. homologous chromosomes move to opposite poles of the cell

_____ 8. chromosomes condense

Name _____ Class _____ Date _____

Active Reading continued

Read the passage below. Then answer the questions that follow.

The stages of meiosis II are as follows:
Prophase II: A new spindle forms around the chromosomes.
Metaphase II: The chromosomes line up along the equator, attached at their centromeres to spindle fibers.
Anaphase II: The centromeres divide, and the chromatids (now called chromosomes) move to opposite poles of the cell.
Telophase II: A nuclear envelope forms around each set of chromosomes. The spindle breaks down, and the cell undergoes cytokinesis. The result of meiosis is four haploid cells.

Match each statement with the stage of meiosis II it describes by writing in the spaces provided, *PII* to represent Prophase II, *MII* to represent Metaphase II, *AII* to represent Anaphase II, or *TII* to represent Telophase II.

_____ 9. centromeres divide

_____ 10. new spindle forms

_____ 11. cell undergoes cytokinesis

_____ 12. chromosomes line up at equator

_____ 13. spindle breaks down

_____ 14. chromosomes move to opposite poles of the cell

_____ 15. four haploid cells form

In the space provided, write the letter of the term or phrase that best completes the statement.

_____ 16. Between meiosis I and meiosis II, chromosomes do NOT
   a. replicate.
   b. change position.
   c. divide.
   d. Both (a) and (b)

Name _____ Class _____ Date _____

Skills Worksheet
Active Reading

Section: Sexual Reproduction

Read the passage below. Then answer the questions that follow.

Some organisms look exactly like their parents and siblings. Others share traits with family members but are not identical to them. Some organisms have two parents, while others have one. The type of reproduction that produces an organism determines how similar the organism is to its parents and siblings. Reproduction, the process of producing offspring, can be asexual or sexual.

In **asexual reproduction,** a single parent passes copies of all its genes to each of its offspring; there is no fusion of haploid cells such as gametes. An individual produced by asexual reproduction is a **clone,** an organism that is genetically identical to its parent. As you have read, prokaryotes reproduce by a type of asexual reproduction called binary fission. Many eukaryotes also produce asexually.

In contrast, in **sexual reproduction,** two parents each form reproductive cells that have one-half the number of chromosomes. A diploid mother and father would give rise to haploid gametes, which join to form diploid offspring. Because both parents contribute genetic material, the offspring have traits of both parents but are not exactly like either parent. Sexual reproduction, with the formation of haploid cells, occurs in eukaryotic organisms, including humans.

SKILL: READING EFFECTIVELY

Read each question, and write your answer in the space provided.

1. Write a sentence that states the main idea of this passage.

   _____

   _____

2. What is a clone?

   _____

   _____

3. What is one form of asexual reproduction?

   _____

   _____

Copyright © by Holt, Rinehart and Winston. All rights reserved.

Holt Biology   Meiosis and Sexual Reproduction

Name _____ Class _____ Date _____

Active Reading continued

4. Why do offspring produced through sexual reproduction show traits of each parent?

_____

_____

5. How are sexual and asexual reproduction alike?

_____

_____

6. How are sexual and asexual reproduction different?

_____

_____

An analogy is a comparison. In the space provided, write the letter of the term that best completes the analogy.

_____ **7.** Asexual reproduction is to one as sexual reproduction is to
  a. many.
  b. fission.
  c. two.
  d. four.

Name _____ Class _____ Date _____

Skills Worksheet

Vocabulary Review

Complete the crossword puzzle using the clues provided.

ACROSS

6. gamete-producing process that occurs in male reproductive organs
7. the kind of reproduction in which two parents form haploid cells that join to produce offspring
9. female gamete
11. occurs during prophase I of meiosis
12. a haploid plant reproductive cell produced by meiosis
13. form of cell division that halves the number of chromosomes
14. the type of assortment that involves the random distribution of homologous chromosomes during meiosis

DOWN

1. the haploid phase of a plant that produces gametes by mitosis
2. the process in most animals that produces diploid zygotes
3. an individual produced by asexual reproduction
4. the kind of reproduction in which a single parent passes copies of all its genes to its offspring
5. the name for the cycle that spans from one generation to the next
7. male gametes
8. diploid phase of a plant that produces spores
10. occurs in the ovaries

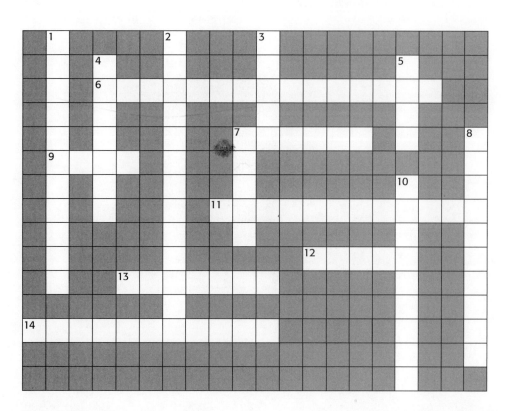

Copyright © by Holt, Rinehart and Winston. All rights reserved.

Holt Biology — Meiosis and Sexual Reproduction

Name _____ Class _____ Date _____

Skills Worksheet

Science Skills

SEQUENCING/ORGANIZING INFORMATION

In the space provided in the figure below, write the letter of the stage of meiosis from the list below (a–h) that matches each stage in the figure.

   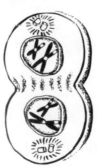

1. _____ 2. _____ 3. _____ 4. _____

  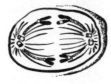 

5. _____ 6. _____ 7. _____ 8. _____

Stages of Meiosis

a. anaphase II   **e.** telophase II and cytokinesis

b. metaphase I   **f.** telophase I and cytokinesis

c. anaphase I   **g.** prophase I

d. metaphase II   **h.** prophase II

Science Skills continued

In the space provided, write the letter of the description that best matches the stage of meiosis.

_____ 9. metaphase I

_____ 10. prophase II

_____ 11. telophase I

_____ 12. metaphase II

_____ 13. telophase II

_____ 14. anaphase II

_____ 15. prophase I

_____ 16. anaphase I

a. A new spindle forms around the chromosomes.

b. Chromatids remain attached at their centromeres as the spindle fibers move the homologous chromosomes to opposite poles of the cell.

c. A nuclear envelope forms around each set of chromosomes, the spindle breaks down, and the cytoplasm divides, resulting in four haploid cells.

d. Chromosomes gather at the poles; the cytoplasm divides.

e. The nuclear envelope breaks down; genetic material is exchanged through crossing-over.

f. Chromosomes line up at the equator.

g. Pairs of homologous chromosomes line up at the equator.

h. Centromeres divide, enabling the chromatids, now called chromosomes, to move to opposite poles of the cell.

Name _____ Class _____ Date _____

Skills Worksheet

Concept Mapping

Using the terms and phrases provided below, complete the concept map showing the process of meiosis.

    chromatids   homologous chromosomes

    crossing-over   meiosis II

    haploid reproductive cells

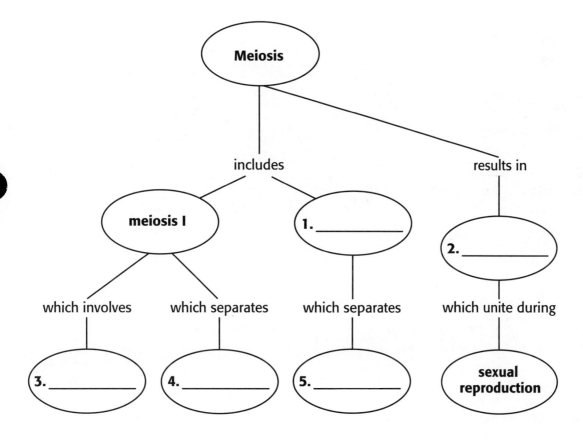

Copyright © by Holt, Rinehart and Winston. All rights reserved.

Holt Biology   Meiosis and Sexual Reproduction

Name _____ Class _____ Date _____

Skills Worksheet

Critical Thinking

Work-Alikes

In the space provided, write the letter of the term or phrase that best describes how each numbered item functions.

_____ 1. anaphase I

_____ 2. crossing-over

_____ 3. metaphase I

_____ 4. meiosis II

_____ 5. telophase I

_____ 6. prophase I

_____ 7. independent assortment

_____ 8. random fertilization

a. bees returning to the hive

b. a meeting

c. shuffling a deck of cards

d. raindrops joining together

e. other half of a baseball inning

f. changing dance partners

g. pulling apart string cheese

h. breaking a container and water spills

Cause and Effect

In the space provided, write the letter of the term or phrase that best matches each cause or effect given below.

Cause	Effect	
9. _____	first cell of new individual	a. gametes
10. meiosis I and II	_____	b. eggs and sperm join
11. _____	gametes join by fusion.	c. all cells of organisms are haploid cells
12. repair of damaged DNA	_____	d. possible explanation of evolution of sexual reproduction

Copyright © by Holt, Rinehart and Winston. All rights reserved.

Holt Biology   17   Meiosis and Sexual Reproduction

Critical Thinking continued

Trade-offs

In the space provided, write the letter of the bad news item that best matches each numbered good news item below.

Good News

_____ 13. Agricultural breeding increases the sizes of domesticated animals.

_____ 14. The ovum produced by oogenesis is very large.

_____ 15. Asexual reproduction allows organisms to reproduce quickly.

_____ 16. Sexual reproduction makes different combinations of genes quickly.

Bad News

a. Organisms must use energy to produce gametes and find mates.

b. The ability to obtain larger animals slows down.

c. They may be at a disadvantage in a changing environment.

d. Two polar bodies produced die.

Linkages

In the spaces provided, write the letters of the two terms or phrases that are linked together by the term or phrase in the middle. The choices can be placed in any order.

17. _____ meiosis II _____

18. _____ independent assortment _____

19. _____ crossing-over _____

20. _____ individuals split off _____

a. budding in hydra

b. four chromatids line up in prophase I

c. gametes

d. meiosis I

e. asexual reproduction

f. recombination of genes in gametes

g. unpaired chromosomes in gametes

h. pairs of chromosomes in organisms

Critical Thinking continued

Analogies

An analogy is a relationship between two pairs of terms or phrases written as a : b :: c : d. The symbol : is read as "is to," and the symbol :: is read as "as." In the space provided, write the letter of the pair of terms or phrases that best completes the analogy shown.

_____ 21. chromatids : crossing-over ::
   a. gametes : chromosomes
   b. metaphase : anaphase
   c. hook : fishing rod
   d. tires : changing flat tire

_____ 22. asexual reproduction : clone ::
   a. spermatogenesis : ovum
   b. fission : budding
   c. meiosis : gamete
   d. fragmentation : budding

_____ 23. haploid life cycle : fusion ::
   a. diploid life cycle : fusion
   b. diploid life cycle : fertilization
   c. fertilization : gametes
   d. gametes : fertilization

_____ 24. sporophyte : spore ::
   a. sexual reproduction : clone
   b. spore : sporophyte
   c. gamete : spore
   d. gametophyte : gamete

Name _____ Class _____ Date _____

Skills Worksheet

Test Prep Pretest

Complete each statement by writing the correct term or phrase in the space provided.

1. Asexual reproduction limits _____ diversity.

2. Spermatogenesis produces _____ sperm cells.

3. Asexual reproduction methods include _____, fragmentation, and _____.

4. In the haploid life cycle, gametes are produced by _____, and the zygote is produced by _____.

5. When corresponding portions of chromatids on two homologous chromosomes change places, _____-_____ has occurred.

6. Only one ovum is produced by _____.

7. In plants that have alternation of generations, the haploid _____ produces the gametes.

8. Increased genetic variation often increases the rate of _____.

9. Meiosis in plants often produces _____, haploid cells that later lead to the production of gametes.

10. Crossing-over is an efficient way to produce _____ _____, which increases genetic diversity.

Name _____ Class _____ Date _____

Test Prep Pretest continued

Questions 11–14 refer to the figure below.

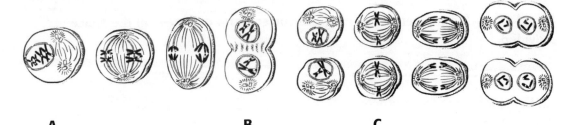

A B C

11. The process shown above is called _____ .

12. In the process shown above, label A refers to _____ .

13. In the process shown above, label B refers to _____ and _____ .

14. In the process shown above, label C refers to _____ .

Read each question, and write your answer in the space provided.

15. Describe the similarities and differences between the formation of male and female gametes.

_____
_____
_____
_____
_____

16. Identify and describe the three types of asexual reproduction.

_____
_____
_____
_____

Name _____ Class _____ Date _____

Test Prep Pretest continued

17. What is the difference between anaphase I and anaphase II? Why is the difference significant?

_____
_____
_____
_____
_____
_____

18. Describe the haploid and diploid life cycles.

_____
_____
_____
_____
_____
_____

19. Describe the advantages and disadvantages of sexual reproduction.

_____
_____
_____
_____
_____

20. How does crossing-over affect evolution?

_____
_____
_____
_____
_____

Name _____ Class _____ Date _____

Assessment

Quiz

Section: Meiosis

In the space provided, write the letter of the term or phrase that best completes each statement or best answers each question.

_____ 1. Crossing-over occurs during
   a. prophase II.   **c.** prophase I.
   b. fertilization.   **d.** metaphase II.

_____ 2. Cytoplasm divides unequally in meiosis during production of
   a. spores.   **c.** cytokinesis.
   b. sperm cells.   **d.** egg cells.

_____ 3. Which of the following does NOT provide new genetic combinations?
   a. random fertilization **c.** independent assortment
   b. cytokinesis   **d.** crossing-over

_____ 4. DNA replication occurs
   a. after telophase I.
   b. prior to prophase I.
   c. in both meiosis I and meiosis II.
   d. when the chromosomes align at the cell's equator.

_____ 5. Spermatogenesis results in
   a. two haploid cells.
   b. three polar bodies.
   c. one haploid sperm cells and three polar bodies.
   d. four haploid sperm cells.

In the space provided, write the letter of the description that best matches the term or phrase.

_____ 6. meiosis   **a.** chromosomes become visible
_____ 7. prophase I   **b.** results in one egg cell and three polar bodies
_____ 8. crossing-over   **c.** results in four haploid cells
_____ 9. telephase II   **d.** halves the number of chromosomes and results in gametes or spores
_____ 10. oogenesis   **e.** results in exchange of chromatid portions between homologous chromosomes

Copyright © by Holt, Rinehart and Winston. All rights reserved.
Holt Biology   Meiosis and Sexual Reproduction

Name _____ Class _____ Date _____

[Assessment]
Quiz

Section: Sexual Reproduction

In the space provided, write the letter of the term or phrase that best completes each statement or best answers each question.

_____ 1. An advantage of sexual reproduction is that
   a. many offspring are produced in a short time.
   b. it increases genetic diversity.
   c. production of gametes requires energy.
   d. organisms remain stable in a changing environment.

_____ 2. The zygote is the only diploid cell in
   a. the haploid life cycle.   c. the diploid life cycle.
   b. asexual reproduction.   d. animals.

_____ 3. Which of the following is NOT a type of asexual reproduction?
   a. budding   c. fission
   b. fragmentation   d. alternation of generations

_____ 4. During an animal's life cycle, if the gametes are the only haploid cells, the life cycle is
   a. alternation of generations. c. a diploid life cycle.
   b. a haploid life cycle.   d. mutated.

_____ 5. Asexual reproduction results in
   a. genetic diversity.
   b. zygotes.
   c. fertilization.
   d. genetically identical offspring.

In the space provided, write the letter of the description that best matches the term or phrase.

_____ 6. fertilization

_____ 7. sporophyte

_____ 8. gametophyte

_____ 9. asexual reproduction

_____ 10. sexual reproduction

Copyright © by Holt, Rinehart and Winston. All rights reserved.
Holt Biology   Meiosis and Sexual Reproduction

Name _____ Class _____ Date _____

Assessment

Chapter Test

Meiosis and Sexual Reproduction

In the space provided, write the letter of the term or phrase that best completes each statement or best answers each question.

_____ 1. Meiosis occurs
   a. in all sexually reproducing organisms.
   b. in all asexually reproducing organisms.
   c. in all reproducing organisms.
   d. during gametogenesis and cytokinesis.

_____ 2. Which of the following is NOT affected by genetic variation?
   a. the development of improved animal breeds
   b. the pace of evolution
   c. the ability of organisms to adapt to changing conditions
   d. the frequency of reproduction

Questions 3 and 4 refer to the figure below, which shows the stages of meiosis.

     A   B   C   D

_____ 3. Pairs of homologous chromosomes line up at the cell's equator in stage
   a. A.
   b. B.
   c. C.
   d. D.

_____ 4. Homologous chromosomes move to opposite poles of the cell during stage
   a. A.
   b. B.
   c. C.
   d. D.

_____ 5. In alternation of generations, which of the following is NOT haploid?
   a. spores
   b. gametophytes
   c. sporophytes
   d. eggs and sperm

Copyright © by Holt, Rinehart and Winston. All rights reserved.

Holt Biology   Meiosis and Sexual Reproduction

Chapter Test continued

_____ 6. During meiosis, two successive divisions
   a. result in the formation of two identical cells.
   b. cause the formation of a zygote.
   c. are responsible for the formation of four haploid cells.
   d. must occur before mitosis can form gametes.

_____ 7. In asexual reproduction,
   a. DNA does not vary much between offspring.
   b. many offspring are produced in a short time.
   c. organisms may not be able to adapt to new environments.
   d. All of the above

_____ 8. Fragmentation is a form of
   a. fission.
   b. asexual reproduction.
   c. crossing-over.
   d. sexual reproduction.

_____ 9. In humans, each gamete receives
   a. 23 pairs of chromosomes from each parent.
   b. one chromosome from each of 23 pairs.
   c. 46 chromosomes.
   d. 23 homologous chromosomes.

Name _____ Class _____ Date _____

Chapter Test *continued*

In the space provided, write the letter of the description that best matches the term or phrase.

_____ 10. crossing-over

_____ 11. life cycle

_____ 12. clone

_____ 13. independent assortment

_____ 14. spore

_____ 15. spermatogenesis

_____ 16. polar body

_____ 17. oogenesis

_____ 18. fertilization

_____ 19. asexual reproduction

_____ 20. ovum

a. random distribution of homologous chromosomes during meiosis

b. small cell with very little cytoplasm that is formed during oogenesis and eventually dies

c. all copies of the single parent's genes are passed to the offspring

d. portions of a chromatid on one homologous chromosome break off and trade places with the corresponding portion on one of the chromatids of the other homologous chromosome

e. the process by which gametes are produced in male animals

f. the union of sperm and egg cells to produce a diploid zygote

g. the activities in the life of an organism from one generation to the next

h. haploid reproductive cell of plants

i. offspring that is genetically identical to its parent

j. female gamete, also called an egg

k. the process by which gametes are produced in female animals

Assessment
Chapter Test

Meiosis and Sexual Reproduction

In the space provided, write the letter of the term or phrase that best completes each statement or best answers each question.

_____ 1. Which of the following is an advantage of sexual reproduction?
   a. The offspring are all genetically identical.
   b. It increases genetic diversity.
   c. It takes energy to find a mate and produce gametes.
   d. Many offspring are produced in a short time.

_____ 2. In alternation of generations, which of the following is haploid?
   a. spores
   b. eggs
   c. sperm
   d. All of the above

_____ 3. In telophase II, cytokinesis results in
   a. two haploid cells.
   b. two diploid cells.
   c. four haploid cells.
   d. four diploid cells.

_____ 4. The final cells resulting from meiosis in either males or females are called
   a. gametes.
   b. polar bodies.
   c. spores.
   d. diploid cells.

_____ 5. During meiosis, the chromatids remain attached at their centromeres until
   a. metaphase I.
   b. cytokinesis.
   c. anaphase I.
   d. anaphase II.

_____ 6. In alternation of generations, spores are produced by the process of
   a. meiosis.
   b. cytokinesis.
   c. mitosis.
   d. spermatogenesis.

Read each question, and write your answer in the space provided.

7. Explain why crossing-over is an important source of genetic variation.

_____

_____

8. Explain how independent assortment and crossing-over can produce a practically unlimited number of genetic combinations among gametes.

_____

_____

Copyright © by Holt, Rinehart and Winston. All rights reserved.
Holt Biology — Meiosis and Sexual Reproduction

Name _____ Class _____ Date _____

Chapter Test continued

9. What is a disadvantage of sexual reproduction?

   _____
   _____
   _____
   _____

10. Explain what happens during alternation of generations in plants.

    _____
    _____
    _____
    _____

11. Compare the processes of spermatogenesis with those of oogenesis.

    _____
    _____
    _____
    _____

12. Explain the difference between asexual reproduction and sexual reproduction.

    _____
    _____
    _____
    _____

Chapter Test continued

In the space provided, write the letter of the description that best matches the term or phrase.

_____ 13. sexual reproduction

_____ 14. prophase I

_____ 15. fragmentation

_____ 16. asexual reproduction

_____ 17. anaphase I

_____ 18. fission

_____ 19. budding

_____ 20. sporophyte

_____ 21. cytokinesis

_____ 22. haploid life cycle

_____ 23. gametophyte

_____ 24. alternation of generations

_____ 25. diploid life cycle

a. separation of a parent into two or more individuals of about equal size

b. produces gametes in the haploid phase of a plant's life cycle

c. haploid cells occupy the major portion of this kind of life cycle

d. a method of asexual reproduction in which the body breaks into several pieces

e. a diploid individual dominates this kind of life cycle

f. produces spores in the diploid phase of a plant's life cycle

g. a life cycle that regularly alternates between a haploid phase and a diploid phase

h. phase of meiosis during which crossing-over occurs

i. all copies of the single parent's genes are passed to the offspring

j. process in which a cell's cytoplasm divides

k. new individuals split off from existing ones

l. homologous chromosomes move to opposite poles of the cell

m. two haploid cells join to form a diploid offspring

Name _____ Class _____ Date _____

Quick Lab

DATASHEET FOR IN-TEXT LAB

Modeling Crossing-Over

You can use paper strips and pencils to model the process of crossing-over.

Homologous chromosomes

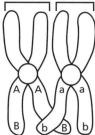

MATERIALS
- 4 paper strips
- pens or pencils (two colors)
- scissors
- tape

Procedure

1. Using one color, write the letters *A* and *B* on two paper strips. These two strips will represent one of the two homologous chromosomes shown above.

2. Using a second color, write the letters *a* and *b* on two paper strips. These two strips will represent the second homologous chromosome shown above.

3. Use your chromosome models, scissors, and tape to demonstrate crossing-over between the chromatids of two homologous chromosomes.

Analysis

1. **Determine** what the letters *A*, *B*, *a*, and *b* represent.

   _____

2. **Infer** why the chromosomes you made are homologous.

   _____

   _____

3. **Compare** the number of different types of chromatids (combinations of *A*, *B*, *a*, and *b*) before crossing-over with the number after crossing-over.

   _____

   _____

4. **Critical Thinking**
   Applying Information How does crossing-over relate to genetic recombination?

   _____

   _____

Name _____ Class _____ Date _____

| Quick Lab | **DATASHEET FOR IN-TEXT LAB** |

Observing Reproduction in Yeast

Yeast are unicellular organisms that live in liquid or moist environments. You can examine a culture of yeast to observe one of the types of reproduction that yeast can undergo.

MATERIALS
- microscope
- microscope slides
- dropper
- culture of yeast

Procedure
1. Make a wet mount of a drop of yeast culture.
2. Observe the yeast with a compound microscope under low power.
3. Look for yeast that appear to be in "pairs."
4. Observe the pairs under high power, and then make drawings of your observations.

Analysis
1. **Infer** the type of reproduction you observed when the yeast appeared to be in pairs.

   _____

   _____

2. **Identify** the reason for your answer.

   _____

   _____

   _____

3. **Determine,** by referring to your textbook, the name of the type of reproduction you observed.

   _____

   _____

   _____

Name _____ Class _____ Date _____

Exploration Lab

DATASHEET FOR IN-TEXT LAB

Modeling Meiosis

SKILLS
- Modeling
- Using scientific methods

OBJECTIVES
- **Describe** the events that occur in each stage of the process of meiosis.
- **Relate** the process of meiosis to genetic variation.

MATERIALS
- pipe cleaners of at least two different colors
- yarn
- wooden beads
- white labels
- scissors

Before You Begin

Meiosis is the process that results in the production of cells with half the normal number of chromosomes. It occurs in all organisms that undergo **sexual reproduction.** In this lab, you will build a model that will help you understand the events of meiosis. You can also use the model to demonstrate the effects of events such as **crossing-over** to explain results such as **genetic recombination.**

1. Write a definition for each boldface term in the paragraph above and for the following terms: homologous chromosomes, gamete. Use a separate sheet of paper.

2. In what organs in the human body do cells undergo meiosis?

   _____

3. During interphase of the cell cycle, how does a cell prepare for dividing?

   _____

   _____

4. Based on the objectives for this lab, write a question you would like to explore about meiosis.

   _____

   _____

Copyright © by Holt, Rinehart and Winston. All rights reserved.

Holt Biology   Meiosis and Sexual Reproduction

Name _____ Class _____ Date _____

Modeling Meiosis continued

Procedure

PART A: DESIGN A MODEL

1. Work with the members of your lab group to design a model of a cell using the materials listed for this lab. Be sure that your model cell has at least two pairs of chromosomes.

2. Use a separate sheet of paper to write out the plan for building your model. Have your teacher approve the plan before you begin building the model.

3. Build the cell model your group designed. **CAUTION: Sharp or pointed objects can cause injury. Handle scissors carefully.** Use your model to demonstrate the phases of meiosis. Draw and label each phase you model.

4. Use your model to explore one of the questions written by your group for step 4 of **Before You Begin.** Describe the steps you took to explore your question.

> **You Choose**
> As you design your experiment, decide the following:
>   a. what question you will explore
>   b. how to construct a cell membrane
>   c. how to show that your cell is diploid
>   d. how to show the locations of at least two genes on each chromosome
>   e. how to show that chromosomes are duplicated before meiosis begins

PART B: TEST HYPOTHESES

Answer each of the following questions by writing a hypothesis. Use your model to test each hypothesis, and describe your results.

5. In humans, gametes (eggs and sperm) result from meiosis. Will all gametes produced by one parent be identical?

_____

_____

_____

6. When an egg and a sperm fuse during sexual reproduction, the resulting cell (the first cell of a new organism) is called a zygote. How many copies of each chromosome and each gene will be found in a zygote?

_____

_____

_____

_____

Name _____ Class _____ Date _____

Modeling Meiosis *continued*

7. Crossing-over frequently occurs between the chromatids of homologous chromosomes during meiosis. Under what circumstances does crossing-over result in new combinations of genes in gametes?

_____

_____

_____

8. Synapsis (the pairing of homologous chromosomes) must occur before crossing-over can take place. How would the outcome of meiosis be different if synapsis did not occur?

_____

_____

_____

PART C: CLEANUP AND DISPOSAL

9. Dispose of paper and yarn scraps in the designated waste container.

10. Clean up your work area and all lab equipment. Return lab equipment to its proper place. Wash your hands thoroughly before you leave the lab and after finishing all work.

Analyze and Conclude

1. **Analyzing Results** How do the nuclei you made by modeling meiosis compare with the nucleus of the cell you started with? Explain your result.

_____

_____

2. **Recognizing Relationships** How are homologous chromosomes different from chromatids?

_____

_____

3. **Forming Reasoned Opinions** How is synapsis important to the outcome of meiosis? Explain.

_____

_____

Copyright © by Holt, Rinehart and Winston. All rights reserved.

Holt Biology — Meiosis and Sexual Reproduction

Name _____ Class _____ Date _____

Modeling Meiosis continued

4. Evaluating Methods How could you modify your model to better illustrate the process of meiosis?

_____

_____

_____

5. Drawing Conclusions How are the processes of meiosis similar to those of mitosis? How are they different?

_____

_____

_____

_____

_____

6. Predicting Outcomes What would happen to the chromosome number of an organism's offspring if the gametes for sexual reproduction were made by mitosis instead of by meiosis?

_____

_____

7. Further Inquiry Write a new question about meiosis or sexual reproduction that could be explored with your model.

_____

_____

_____

Name _____ Class _____ Date _____

Exploration Lab **MODELING**

Karyotyping

Humans have 46 chromosomes in every diploid (2n) body cell. The chromosomes of a diploid cell occur in *homologous pairs*, which are pairs of chromosomes that are similar in size, shape, and the position of their centromere. In humans, 22 homologous pairs of chromosomes are called *autosomes*. The twenty-third pair, which determines the individual's sex, make up the *sex chromosomes*. Females have only one type of sex chromosome, which is called an *X chromosome*. Males have two types of sex chromosomes, an X chromosome and a much smaller *Y chromosome*. **Figure 1** on the next page shows each of the 22 types of autosomes and the 2 types of sex chromosomes.

A *karyotype* is a diagram that shows a cell's chromosomes arranged in order from largest to smallest. A karyotype is made from a photomicrograph (photo taken through a microscope) of the chromosomes from a cell in metaphase. The photographic images of the chromosomes are cut out and arranged in homologous pairs by their size and shape. The karyotype can be analyzed to determine the sex of the individual and whether there are any chromosomal abnormalities. For example, the karyotype of a female shows two X chromosomes, and the karyotype of a male shows an X chromosome and a Y chromosome.

Chromosomal abnormalities often result from *nondisjunction*, the failure of chromosomes to separate properly during meiosis. Nondisjunction results in cells that have too many or too few chromosomes. *Trisomy* is an abnormality in which a cell has an extra chromosome, or section of a chromosome. This means that the cell contains 47 chromosomes instead of 46. *Down syndrome*, or *trisomy 21*, is a chromosomal abnormality that results from having an extra number 21 chromosome.

In this lab, you will complete and analyze a karyotype of cells from a fetus to determine the sex of the fetus and whether or not the fetus has Down syndrome.

OBJECTIVES

Identify pairs of homologous chromosomes by their length, centromere position, and banding pattern.

Determine the sex of an individual from a karyotype.

Predict whether or not an individual will have Down syndrome.

MATERIALS

- chromosome spread
- metric ruler
- scissors
- transparent tape
- WARD'S human karyotyping form

Karyotyping continued

FIGURE 1 HUMAN CHROMOSOMES

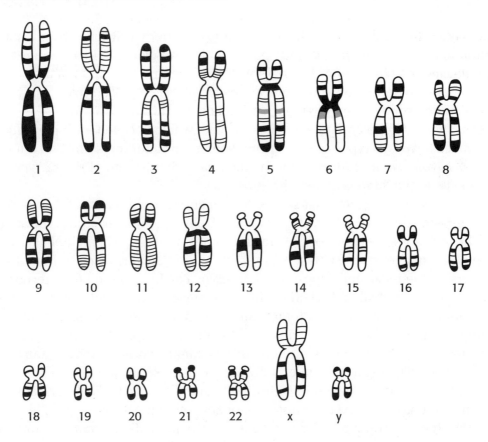

Procedure

1. Carefully cut each chromosome from the chromosome spread in **Figure 2**. Be sure to leave a slight margin around each chromosome.

2. Arrange the chromosomes in homologous pairs. The members of each pair will be the same length and will have the centromere in the same location. Use the ruler to measure the length of the chromosome and the position of the centromere. Arrange the pairs according to their length, from largest to smallest. The banding patterns of the chromosomes may also help you to pair up the homologous chromosomes.

3. Tape each homologous pair to a human karyotyping form, positioning the centromeres on the lines. Place the pairs in order, with the longest pair at position 1, the shortest pair at position 22, and the sex chromosomes at position 23.

4. The diagram you have made is a karyotype. Analyze the karyotype to determine the sex of the individual and whether or not the individual will have Down syndrome.

5. Clean up your materials before leaving the lab.

Name _____ Class _____ Date _____

Karyotyping continued

Analysis

1. **Examining Data** Examine your karyotype. Is the fetus male or female? How do you know?

   _____

   _____

2. **Examining Data** Will the baby have Down syndrome? How do you know?

   _____

   _____

   _____

3. **Explaining Events** The Y chromosome closely resembles many of the other chromosomes. What did you have to do to determine that it was the Y chromosome?

   _____

   _____

   _____

   _____

   _____

4. **Recognizing Patterns** If the karyotype you constructed was for a female with Down syndrome, what chromosome changes would be evident?

   _____

   _____

   _____

5. **Identifying Relationships** A pedigree is a diagram that shows the presence or absence of a trait in each person in each generation. How does a karyotype differ from a pedigree?

   _____

   _____

   _____

   _____

Name _____ Class _____ Date _____

Karyotyping *continued*

Conclusions

1. **Interpreting Information** If your job were to inform the parents of the fetus of their test results, what would you say?

   _____

   _____

   _____

2. **Drawing Conclusions** Why are karyotypes important tools for geneticists?

   _____

   _____

3. **Evaluating Models** Can the analysis of a karyotype reveal point mutations? Explain your answer.

   _____

   _____

4. **Applying Conclusions** You have prepared a karyotype of an individual and have found that one of the chromatids of chromosome 4 is shorter than its homologue. What can you conclude has happened to this chromosome?

   _____

   _____

Extensions

1. **Research and Communications** *Genetic counselors* work with couples who are concerned that the father or mother may have a harmful gene that could be passed to their child. The counselors study the family history and perform blood tests to detect harmful genes. They then predict if there is a chance that a parent could contribute a harmful gene to a child. Parents can then make an informed decision about having children. Find out about the training and skills required to become a genetic counselor.

2. **Research and Communications** Research genetic diseases. Collect data about the incidences of genetic diseases by age group and number of cases. Present your findings with graphs and pictures.

Karyotyping *continued*

FIGURE 2 CHROMOSOME SPREAD

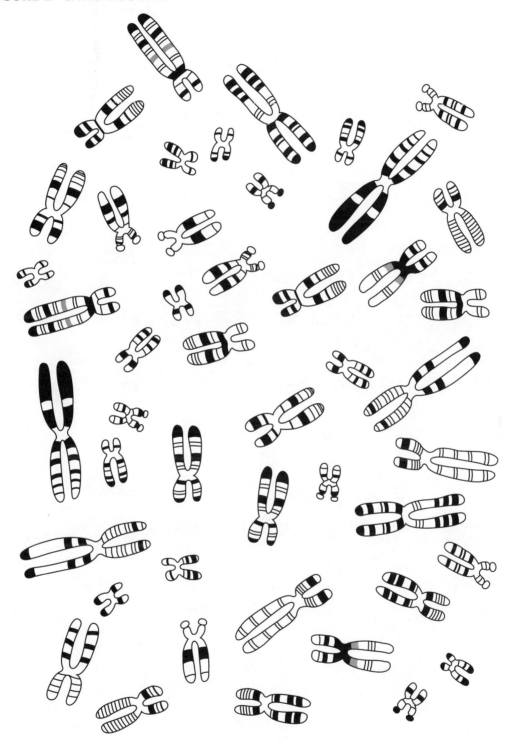

Name _____ Class _____ Date _____

Skills Practice Lab   OBSERVATION

Laboratory Techniques: Staining DNA and RNA

Root tip cells of onions *(Allium sativum)* are frequently used to study DNA and RNA in plant cells. In plants, mitosis occurs in special growth regions called *meristems* located at the tips of the roots and stems. To observe chromosomes in stem and root meristems, biologists prepare a special kind of slide called a *squash*. This preparation is just what it sounds like. Tissue containing actively dividing cells is removed from a root or stem meristem and treated with hydrochloric acid to fix the cells, or to stop them from dividing. The cells are then stained, made into a wet mount, and squashed and spread into a single layer by applying pressure to the coverslip.

The stain methyl green-pyronin Y is a mixture of two different stains. Methyl green is absorbed by DNA only and stains the DNA blue. Pyronin Y is absorbed by RNA only and stains the RNA pink. Therefore, methyl green-pyronin Y can be used to differentiate between the two nucleic acids.

In this lab, you will use methyl green-pyronin Y stain for detecting both DNA and RNA. You will observe the difference in staining results.

OBJECTIVES

Demonstrate how to make a squash of onion root tips.

Compare the location of DNA and RNA in the cell.

MATERIALS

- compound light microscope
- coverslip
- forceps
- lab apron
- safety goggles
- methyl green-pyronin Y stain in dropper bottle
- microscope slide
- paper towels
- pencil with eraser
- prepared reference slide of plant cells
- prepared reference slide of animal cells
- vial of pretreated *Allium* root tips
- watch or clock
- wooden macerating stick

Holt Biology   51   Meiosis and Sexual Reproduction

Name _____ Class _____ Date _____

Laboratory Techniques: Staining DNA and RNA *continued*

Procedure

1. Put on safety goggles and a lab apron.

2. Place a microscope slide on a paper towel on a smooth, flat surface. **CAUTION: Glassware is fragile. Notify your teacher promptly of any broken glass or cuts.** Add two drops of methyl green-pyronin Y stain to the center of the slide. **CAUTION: Methyl green-pyronin Y stain will stain your skin and clothing. Promptly wash off spills to minimize staining.**

3. Use forceps to transfer a prepared onion root tip to the drop of stain on the microscope slide.

4. Carefully smash the root tip by gently but firmly tapping the root with the end of a wooden macerating stick. *Note: Tap the macerating stick in a straight up-and-down motion.*

5. Allow the root tip to stain for 10 to 15 minutes. *Note: Do not let the stain dry. Add more stain if necessary.*

6. Place a coverslip over your preparation, and cover the slide by folding a paper towel over it. Using the eraser end of a pencil, gently, but firmly press straight down (with no twisting) on the coverslip through the paper towel. Apply only enough pressure to squash the root tip into a single cell layer. *Note: Be very careful not to move the coverslip while you are pressing down with the pencil eraser. Also be very careful not to press too hard. If you press too hard, you might break the glass slide or tear apart the cells in the onion root tip.*

7. Examine your prepared slide under both the low power and the high power of a compound light microscope. *Note: Remember that your mount is fairly thick, so be careful not to switch to the high-power objective too quickly. You may shatter the coverslip and destroy your preparation. You will need to focus up and down carefully with the fine adjustment to better see the structures under study.* Record the color of the stain for each structure in **Table 1**.

TABLE 1 COLOR OF CELL STRUCTURES IN ONION ROOT TIP CELLS

Structure	Stained color
Nucleus	
Nucleolus	
Cytoplasm	
Chromosomes	

Holt Biology — Meiosis and Sexual Reproduction

Name _____ Class _____ Date _____

Laboratory Techniques: Staining DNA and RNA *continued*

8. In the space below, draw and label a representative plant cell from your prepared slide. Include all visible organelles. Indicate where DNA and RNA are found in the cell.

9. Observe the prepared reference slides of plant and animal cells.
   - How does your slide compare with the prepared slide of onion root tip cells?

   _____
   _____

10. Dispose of your materials according to the directions from your teacher.

11. Clean up your work area and wash your hands before leaving the lab.

Analysis

1. **Describing Events** What color did the deoxyribonucleic acid (DNA) stain in your root tip squash?

   _____

2. **Analyzing Data** How do you know this material is DNA?

   _____
   _____

Name _____ Class _____ Date _____

Laboratory Techniques: Staining DNA and RNA *continued*

3. Analyzing Results What color did the nucleoli appear in the stained slide? What does this tell you about the composition of nucleoli?

_____

_____

4. Examining Data Did each nucleus have only one nucleolus or several? What appeared to be the most common number of nucleoli?

_____

_____

5. Describing Events Were you able to see any cells in the process of mitotic division? If so, what did the cells look like?

_____

_____

_____

Conclusions

1. Drawing Conclusions Where is DNA located in both plant and animal cells?

_____

_____

2. Drawing Conclusions Where is RNA located in both plant and animal cells?

_____

_____

Extension

Research and Communications Use library references to research other staining techniques.

Name _____ Class _____ Date _____

Quick Lab

DATASHEET FOR IN-TEXT LAB

Modeling Crossing-Over

You can use paper strips and pencils to model the process of crossing-over.

MATERIALS
- 4 paper strips
- pens or pencils (two colors)
- scissors
- tape

Procedure

1. Using one color, write the letters A and B on two paper strips. These two strips will represent one of the two homologous chromosomes shown above.

2. Using a second color, write the letters a and b on two paper strips. These two strips will represent the second homologous chromosome shown above.

3. Use your chromosome models, scissors, and tape to demonstrate crossing-over between the chromatids of two homologous chromosomes.

Analysis

1. **Determine** what the letters A, B, a, and b represent.

   individual genes

2. **Infer** why the chromosomes you made are homologous.

   The chromosomes are of similar size, shape, and genetic content.

3. **Compare** the number of different types of chromatids (combinations of A, B, a, and b) before crossing-over with the number after crossing-over.

   Before crossing-over: *AB, AB, ab,* and *ab*. After crossing-over: answers will

   vary (for example, *AB, Ab, aB,* and *ab*)

4. **Critical Thinking**
   Applying Information How does crossing-over relate to genetic recombination?

   Crossing-over causes genetic recombination.

TEACHER RESOURCE PAGE

Name _____ Class _____ Date _____

Quick Lab

DATASHEET FOR IN-TEXT LAB

Observing Reproduction in Yeast

Yeast are unicellular organisms that live in liquid or moist environments. You can examine a culture of yeast to observe one of the types of reproduction that yeast can undergo.

MATERIALS

- microscope
- microscope slides
- dropper
- culture of yeast

Procedure

1. Make a wet mount of a drop of yeast culture.
2. Observe the yeast with a compound microscope under low power.
3. Look for yeast that appear to be in "pairs."
4. Observe the pairs under high power, and then make drawings of your observations.

Analysis

1. **Infer** the type of reproduction you observed when the yeast appeared to be in pairs.

   asexual reproduction

2. **Identify** the reason for your answer.

   Only one parent was observed. The new organism split off from the parent.

3. **Determine,** by referring to your textbook, the name of the type of reproduction you observed.

   budding

Copyright © by Holt, Rinehart and Winston. All rights reserved.

TEACHER RESOURCE PAGE

Name _____ Class _____ Date _____

Exploration Lab

Modeling Meiosis

DATASHEET FOR IN-TEXT LAB

SKILLS
- Modeling
- Using scientific methods

OBJECTIVES
- **Describe** the events that occur in each stage of the process of meiosis.
- **Relate** the process of meiosis to genetic variation.

MATERIALS
- pipe cleaners of at least two different colors
- yarn
- wooden beads
- white labels
- scissors

Before You Begin

Meiosis is the process that results in the production of cells with half the normal number of chromosomes. It occurs in all organisms that undergo **sexual reproduction**. In this lab, you will build a model that will help you understand the events of meiosis. You can also use the model to demonstrate the effects of events such as **crossing-over** to explain results such as **genetic recombination**.

1. Write a definition for each boldface term in the paragraph above and for the following terms: homologous chromosomes, gamete. Use a separate sheet of paper. **Answers appear in the TE for this lab.**

2. In what organs in the human body do cells undergo meiosis?
   Meiosis occurs in the ovaries and testes.

3. During interphase of the cell cycle, how does a cell prepare for dividing?
   A cell's chromosomes duplicate, and certain organelles replicate.

4. Based on the objectives for this lab, write a question you would like to explore about meiosis.
   Answers will vary. For example: How many chromosomes will each new nucleus have after meiosis has occurred?

Copyright © by Holt, Rinehart and Winston. All rights reserved.

Holt Biology   57   Meiosis and Sexual Reproduction

TEACHER RESOURCE PAGE

Name _____ Class _____ Date _____

Modeling Meiosis *continued*

Procedure
PART A: DESIGN A MODEL

1. Work with the members of your lab group to design a model of a cell using the materials listed for this lab. Be sure that your model cell has at least two pairs of chromosomes.

2. Use a separate sheet of paper to write out the plan for building your model. Have your teacher approve the plan before you begin building the model. **Answers appear in the TE for this lab.**

3. Build the cell model your group designed. **CAUTION: Sharp or pointed objects can cause injury. Handle scissors carefully.** Use your model to demonstrate the phases of meiosis. Draw and label each phase you model.

4. Use your model to explore one of the questions written by your group for step 4 of Before You Begin. Describe the steps you took to explore your question.

> **You Choose**
> As you design your experiment, decide the following:
> a. what question you will explore
> b. how to construct a cell membrane
> c. how to show that your cell is diploid
> d. how to show the locations of at least two genes on each chromosome
> e. how to show that chromosomes are duplicated before meiosis begins

PART B: TEST HYPOTHESES

Answer each of the following questions by writing a hypothesis. Use your model to test each hypothesis, and describe your results.

5. In humans, gametes (eggs and sperm) result from meiosis. Will all gametes produced by one parent be identical?

   Hypotheses will vary. Students should find that the gametes will be identical only if the parent is homozygous for every one of its genetic traits.

6. When an egg and a sperm fuse during sexual reproduction, the resulting cell (the first cell of a new organism) is called a zygote. How many copies of each chromosome and each gene will be found in a zygote?

   Hypotheses will vary. Students should find two copies of each chromosome and two copies of each gene in a zygote.

TEACHER RESOURCE PAGE

Name _____ Class _____ Date _____

Modeling Meiosis continued

7. Crossing-over frequently occurs between the chromatids of homologous chromosomes during meiosis. Under what circumstances does crossing-over result in new combinations of genes in gametes?

   Hypotheses will vary. Students should find that crossing-over can produce

   new combinations of genes when an organism has different versions of the

   genes on the parts that cross over.

8. Synapsis (the pairing of homologous chromosomes) must occur before crossing-over can take place. How would the outcome of meiosis be different if synapsis did not occur?

   Hypotheses will vary. Students may find the wrong chromosome number in a

   gamete if the homologous pairs do not separate properly during anaphase I.

   Also, there would be less genetic variation.

PART C: CLEANUP AND DISPOSAL

9. Dispose of paper and yarn scraps in the designated waste container.

10. Clean up your work area and all lab equipment. Return lab equipment to its proper place. Wash your hands thoroughly before you leave the lab and after finishing all work.

Analyze and Conclude

1. **Analyzing Results** How do the nuclei you made by modeling meiosis compare with the nucleus of the cell you started with? Explain your result.

   The nuclei made by meiosis have half the original chromosome number.

   Two divisions of the nuclear material occur.

2. **Recognizing Relationships** How are homologous chromosomes different from chromatids?

   Homologous chromosomes are the same size and have genes for the same

   traits, but they are not identical, as are chromatids.

3. **Forming Reasoned Opinions** How is synapsis important to the outcome of meiosis? Explain.

   Synapses ensure that each new cell will get one member of each pair of

   homologous chromosomes.

Copyright © by Holt, Rinehart and Winston. All rights reserved.

Holt Biology — Meiosis and Sexual Reproduction

Modeling Meiosis continued

4. Evaluating Methods How could you modify your model to better illustrate the process of meiosis?

Answers will vary. Students may suggest using different materials.

5. Drawing Conclusions How are the processes of meiosis similar to those of mitosis? How are they different?

Mitosis and meiosis are both forms of nuclear division. Both consist of prophase, metaphase, anaphase, and telophase. Mitosis consists of one division and results in two cells that are genetically the same and have the same number of chromosomes as the original cell. Meiosis consists of two divisions and results in four cells that each have half the number of chromosomes as the original cell and are not the same genetically.

6. Predicting Outcomes What would happen to the chromosome number of an organism's offspring if the gametes for sexual reproduction were made by mitosis instead of by meiosis?

The chromosome number would be twice that in its parents' cells.

7. Further Inquiry Write a new question about meiosis or sexual reproduction that could be explored with your model.

Answers will vary. For example: How do mitosis and meiosis compare?

TEACHER RESOURCE PAGE

Exploration Lab

Karyotyping

MODELING

Teacher Notes

TIME REQUIRED One 45-minute period

SKILLS ACQUIRED
Identifying patterns
Inferring
Interpreting
Organizing and analyzing data

RATING Easy  Hard

Teacher Prep–1
Student Setup–1
Concept Level–2
Cleanup–1

THE SCIENTIFIC METHOD

Make Observations Students observe chromosomes to complete and analyze a karyotype.

Analyze the Results Analysis questions 1–5 require students to analyze their results.

Draw Conclusions Conclusions questions 1–4 ask students to draw conclusions from their data.

MATERIALS

Materials for this lab can be purchased from WARD'S. See the *Master Materials List* for ordering instructions.

SAFETY CAUTIONS

Discuss all safety symbols with students

TIPS AND TRICKS
Preparation

- If a karyotype form is not available, students can make their own. Be sure to have them label groups A–G and place the centromeres on the line. They also need to number them accordingly.

Copyright © by Holt, Rinehart and Winston. All rights reserved.

Holt Biology   Meiosis and Sexual Reproduction

Karyotyping continued

Procedure

- Instruct students to cut out the chromosomes in a rectangular shape. It is a good idea to place the cut-out chromosomes face up in one pile and the scrap paper in another pile.
- When taping down the chromosomes, students will find it easier to use short pieces of tape and to tape down one or two pairs at a time. Students are likely to make a mistake if they try to tape down an entire row at one time.
- When evaluating the karyotype, check that chromosome pairs are accurate and that each pair is centered on the line by their centromeres.

Additional Background

You may want to discuss genetic diseases such as Down syndrome. When discussing nondisjunction, you may want to introduce the following terms:

- deletion—a portion of a chromosome is lost
- duplication—the deletion becomes incorporated into its homologue so that the segment appears twice on the same chromosome
- translocation—the deleted portion is transferred to a chromosome other than its homologue

TEACHER RESOURCE PAGE

Name _____ Class _____ Date _____

Exploration Lab   **MODELING**
Karyotyping

Humans have 46 chromosomes in every diploid ($2n$) body cell. The chromosomes of a diploid cell occur in *homologous pairs*, which are pairs of chromosomes that are similar in size, shape, and the position of their centromere. In humans, 22 homologous pairs of chromosomes are called *autosomes*. The twenty-third pair, which determines the individual's sex, make up the *sex chromosomes*. Females have only one type of sex chromosome, which is called an X *chromosome*. Males have two types of sex chromosomes, an X chromosome and a much smaller *Y chromosome*. **Figure 1** on the next page shows each of the 22 types of autosomes and the 2 types of sex chromosomes.

A *karyotype* is a diagram that shows a cell's chromosomes arranged in order from largest to smallest. A karyotype is made from a photomicrograph (photo taken through a microscope) of the chromosomes from a cell in metaphase. The photographic images of the chromosomes are cut out and arranged in homologous pairs by their size and shape. The karyotype can be analyzed to determine the sex of the individual and whether there are any chromosomal abnormalities. For example, the karyotype of a female shows two X chromosomes, and the karyotype of a male shows an X chromosome and a Y chromosome.

Chromosomal abnormalities often result from *nondisjunction*, the failure of chromosomes to separate properly during meiosis. Nondisjunction results in cells that have too many or too few chromosomes. *Trisomy* is an abnormality in which a cell has an extra chromosome, or section of a chromosome. This means that the cell contains 47 chromosomes instead of 46. *Down syndrome*, or *trisomy 21*, is a chromosomal abnormality that results from having an extra number 21 chromosome.

In this lab, you will complete and analyze a karyotype of cells from a fetus to determine the sex of the fetus and whether or not the fetus has Down syndrome.

OBJECTIVES

Identify pairs of homologous chromosomes by their length, centromere position, and banding pattern.

Determine the sex of an individual from a karyotype.

Predict whether or not an individual will have Down syndrome.

MATERIALS

- chromosome spread
- metric ruler
- scissors
- transparent tape
- WARD'S human karyotyping form

Copyright © by Holt, Rinehart and Winston. All rights reserved.
Holt Biology   Meiosis and Sexual Reproduction

Name _____ Class _____ Date _____

Karyotyping *continued*

FIGURE 1 HUMAN CHROMOSOMES

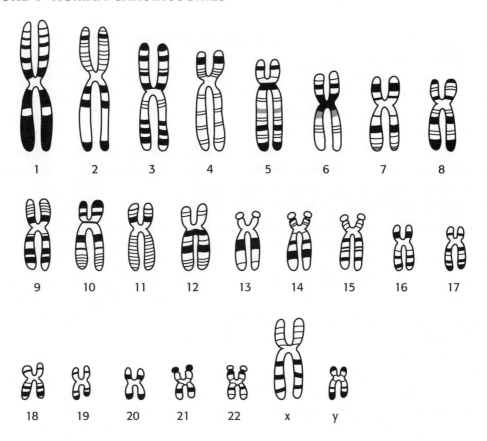

Procedure

1. Carefully cut each chromosome from the chromosome spread in **Figure 2**. Be sure to leave a slight margin around each chromosome.

2. Arrange the chromosomes in homologous pairs. The members of each pair will be the same length and will have the centromere in the same location. Use the ruler to measure the length of the chromosome and the position of the centromere. Arrange the pairs according to their length, from largest to smallest. The banding patterns of the chromosomes may also help you to pair up the homologous chromosomes.

3. Tape each homologous pair to a human karyotyping form, positioning the centromeres on the lines. Place the pairs in order, with the longest pair at position 1, the shortest pair at position 22, and the sex chromosomes at position 23.

4. The diagram you have made is a karyotype. Analyze the karyotype to determine the sex of the individual and whether or not the individual will have Down syndrome.

5. Clean up your materials before leaving the lab.

Name _____ Class _____ Date _____

Karyotyping *continued*

Analysis

1. **Examining Data** Examine your karyotype. Is the fetus male or female? How do you know?

    The baby is a normal male. The sex is determined by examining the sex

    chromosomes. There is an X and a Y.

2. **Examining Data** Will the baby have Down syndrome? How do you know?

    The autosomes are all paired and there are no extras, so the karyotype is

    normal. The baby will not have Down syndrome.

3. **Explaining Events** The Y chromosome closely resembles many of the other chromosomes. What did you have to do to determine that it was the Y chromosome?

    The length of the chromosome and position of the centromere were meas-

    ured. Banding patterns were compared, and it was determined that there was

    only one chromosome present and not a pair. This helped to distinguish the

    Y chromosome from others of similar size and shape.

4. **Recognizing Patterns** If the karyotype you constructed was for a female with Down syndrome, what chromosome changes would be evident?

    There would be two X chromosomes indicating a female, and an extra 21st

    chromosome indicating trisomy 21, or Down syndrome. The total chromo-

    some number would increase to 47.

5. **Identifying Relationships** A pedigree is a diagram that shows the presence or absence of a trait in each person in each generation. How does a karyotype differ from a pedigree?

    A karyotype provides information about a single individual, and does not

    show several generations. A karyotype is a diagram of an individual's chro-

    mosomes that can show chromosomal mutations. A pedigree traces a trait or

    point mutation over several generations.

TEACHER RESOURCE PAGE

Name _____ Class _____ Date _____

Karyotyping *continued*

Conclusions

1. **Interpreting Information** If your job were to inform the parents of the fetus of their test results, what would you say?

   Answers will vary. Students should cite evidence to support their conclusions that the fetus is male, and does not have Down syndrome. Answers should be specific to gender and Down syndrome. Other conditions may exist that cannot be determined by karyotype.

2. **Drawing Conclusions** Why are karyotypes important tools for geneticists?

   Karyotypes supply important, specific information related to certain genetic chromosomal problems and the sex of an individual.

3. **Evaluating Models** Can the analysis of a karyotype reveal point mutations? Explain your answer.

   No. Karyotypes can only reveal certain chromosomal mutations. Point mutations involve base pairs in the DNA, and these cannot be seen in a karyotype.

4. **Applying Conclusions** You have prepared a karyotype of an individual and have found that one of the chromatids of chromosome 4 is shorter than its homologue. What can you conclude has happened to this chromosome?

   Chromosome 4 has undergone a deletion in which part of the chromosome has been lost.

Extensions

1. **Research and Communications** *Genetic counselors* work with couples who are concerned that the father or mother may have a harmful gene that could be passed to their child. The counselors study the family history and perform blood tests to detect harmful genes. They then predict if there is a chance that a parent could contribute a harmful gene to a child. Parents can then make an informed decision about having children. Find out about the training and skills required to become a genetic counselor.

2. **Research and Communications** Research genetic diseases. Collect data about the incidences of genetic diseases by age group and number of cases. Present your findings with graphs and pictures.

Copyright © by Holt, Rinehart and Winston. All rights reserved.

Holt Biology　　　　　　　　　　Meiosis and Sexual Reproduction

Name _____ Class _____ Date _____

Karyotyping continued

FIGURE 2 CHROMOSOME SPREAD

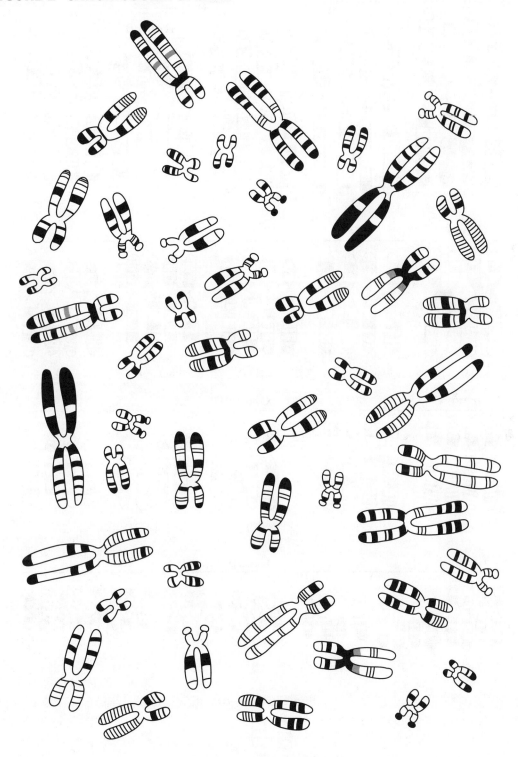

Karyotyping continued

Human Karyotypes Answer Key

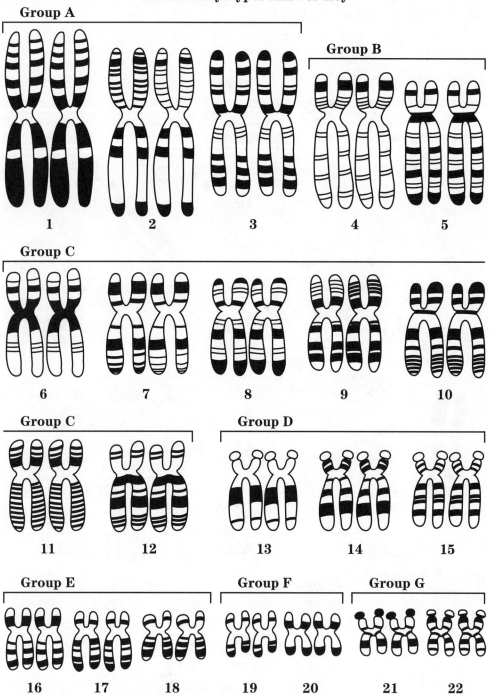

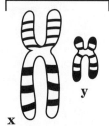

TEACHER RESOURCE PAGE

Name _____ Class _____ Date _____

Skills Practice Lab   OBSERVATION

Laboratory Techniques: Staining DNA and RNA

Teacher Notes

TIME REQUIRED One 45-minute period

SKILLS ACQUIRED
Collecting data
Identifying patterns
Inferring
Interpreting
Organizing and analyzing data

RATING
Teacher Prep–2 Easy ←1—2—3—4→ Hard
Student Setup–2
Concept Level–3
Cleanup–2

THE SCIENTIFIC METHOD

Make Observations Students observe stains made for detecting DNA and RNA.

Analyze the Results Analysis questions 1–5 require students to analyze their results.

Draw Conclusions Conclusions questions 1 and 2 require students to draw conclusions from their data.

MATERIALS

Materials for this lab can be purchased from WARD'S. See the *Master Materials List* for ordering instructions.

SAFETY CAUTIONS

- Discuss all safety symbols and caution statements with students.
- Review the rules for carrying and using the compound microscope.
- Remind students to notify you immediately of any chemical spills. Also caution them to never taste, touch, or smell any substance or bring it close to their eyes.
- Methyl green-pyronin Y stain is an irritant. Avoid skin and eye contact. It can stain skin and clothing. In case of contact, flush affected areas with water for 15 minutes, including under the eyelids; rinse mouth with water.

Copyright © by Holt, Rinehart and Winston. All rights reserved.
Holt Biology   Meiosis and Sexual Reproduction

Name _____ Class _____ Date _____

Laboratory Techniques: Staining DNA and RNA *continued*

DISPOSAL

Methyl green-pyronin Y is a water-based stain. It can be washed down the drain with water.

TIPS AND TRICKS
Procedure

- Students should be able to locate DNA (stained blue-purple) and RNA (stained pink-red), as well as a nucleus, multiple nucleoli, and granular cytoplasmic bodies, in the onion cells. Students may also be able to locate granular chromatin and condensed chromosomes in cells in the process of division.

- The onion root tip preparation the students make may be fairly thick. Caution them to take care not to force the objective down onto the slide they prepare, either crushing the slide or damaging the high-power objective.

- Place prepared reference slides of DNA/RNA in animal cells and nucleic acids in plant cells, in focus, under low power on a class-accessible microscope. If possible, project the slides on a projecting scope.

Additional Background

Nucleic acids, along with proteins, fats, and carbohydrates, are the four major groups of organic chemicals that make up cells in an organism. Nucleic acids are responsible for storing information about the structure of proteins and the genetic makeup of an organism. Nucleic acids also control the reproduction of cells.

A nucleic acid contains only five elements: carbon, hydrogen, oxygen, nitrogen, and phosphorus. These elements combine in three distinct subunits to form a nucleotide, the smallest piece of a nucleic acid strand. Nucleotides contain one of two sugars. Ribose is found only in ribonucleic acid (RNA). Deoxyribose is found only in deoxyribose nucleic acids (DNA).

Name _____ Class _____ Date _____

Skills Practice Lab   OBSERVATION

Laboratory Techniques: Staining DNA and RNA

Root tip cells of onions *(Allium sativum)* are frequently used to study DNA and RNA in plant cells. In plants, mitosis occurs in special growth regions called *meristems* located at the tips of the roots and stems. To observe chromosomes in stem and root meristems, biologists prepare a special kind of slide called a *squash*. This preparation is just what it sounds like. Tissue containing actively dividing cells is removed from a root or stem meristem and treated with hydrochloric acid to fix the cells, or to stop them from dividing. The cells are then stained, made into a wet mount, and squashed and spread into a single layer by applying pressure to the coverslip.

The stain methyl green-pyronin Y is a mixture of two different stains. Methyl green is absorbed by DNA only and stains the DNA blue. Pyronin Y is absorbed by RNA only and stains the RNA pink. Therefore, methyl green-pyronin Y can be used to differentiate between the two nucleic acids.

In this lab, you will use methyl green-pyronin Y stain for detecting both DNA and RNA. You will observe the difference in staining results.

OBJECTIVES

Demonstrate how to make a squash of onion root tips.

Compare the location of DNA and RNA in the cell.

MATERIALS

- compound light microscope
- coverslip
- forceps
- lab apron
- safety goggles
- methyl green-pyronin Y stain in dropper bottle
- microscope slide
- paper towels
- pencil with eraser
- prepared reference slide of plant cells
- prepared reference slide of animal cells
- vial of pretreated *Allium* root tips
- watch or clock
- wooden macerating stick

Holt Biology   Meiosis and Sexual Reproduction

Name _____ Class _____ Date _____

Laboratory Techniques: Staining DNA and RNA *continued*

Procedure

1. Put on safety goggles and a lab apron.

2. Place a microscope slide on a paper towel on a smooth, flat surface. **CAUTION: Glassware is fragile. Notify your teacher promptly of any broken glass or cuts.** Add two drops of methyl green-pyronin Y stain to the center of the slide. **CAUTION: Methyl green-pyronin Y stain will stain your skin and clothing. Promptly wash off spills to minimize staining.**

3. Use forceps to transfer a prepared onion root tip to the drop of stain on the microscope slide.

4. Carefully smash the root tip by gently but firmly tapping the root with the end of a wooden macerating stick. *Note: Tap the macerating stick in a straight up-and-down motion.*

5. Allow the root tip to stain for 10 to 15 minutes. *Note: Do not let the stain dry. Add more stain if necessary.*

6. Place a coverslip over your preparation, and cover the slide by folding a paper towel over it. Using the eraser end of a pencil, gently, but firmly press straight down (with no twisting) on the coverslip through the paper towel. Apply only enough pressure to squash the root tip into a single cell layer. *Note: Be very careful not to move the coverslip while you are pressing down with the pencil eraser. Also be very careful not to press too hard. If you press too hard, you might break the glass slide or tear apart the cells in the onion root tip.*

7. Examine your prepared slide under both the low power and the high power of a compound light microscope. *Note: Remember that your mount is fairly thick, so be careful not to switch to the high-power objective too quickly. You may shatter the coverslip and destroy your preparation. You will need to focus up and down carefully with the fine adjustment to better see the structures under study.* Record the color of the stain for each structure in **Table 1.**

TABLE 1 COLOR OF CELL STRUCTURES IN ONION ROOT TIP CELLS

Structure	Stained color
Nucleus	dark reddish-purple
Nucleolus	red, pink
Cytoplasm	light pink
Chromosomes	blue, purple

Laboratory Techniques: Staining DNA and RNA *continued*

8. In the space below, draw and label a representative plant cell from your prepared slide. Include all visible organelles. Indicate where DNA and RNA are found in the cell.

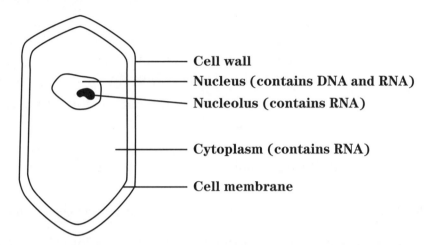

9. Observe the prepared reference slides of plant and animal cells.

- How does your slide compare with the prepared slide of onion root tip cells?

  Answers will vary depending on students' technique.

10. Dispose of your materials according to the directions from your teacher.

11. Clean up your work area and wash your hands before leaving the lab.

Analysis

1. **Describing Events** What color did the deoxyribonucleic acid (DNA) stain in your root tip squash?

   Methyl green-pyronin Y stains DNA blue to purple.

2. **Analyzing Data** How do you know this material is DNA?

   It is located in the nucleus, but not within the nucleolus, where RNA is found.

Name _____ Class _____ Date _____

Laboratory Techniques: Staining DNA and RNA *continued*

3. Analyzing Results What color did the nucleoli appear in the stained slide? What does this tell you about the composition of nucleoli?

Methyl green-pyronin Y stains nucleoli red to pink. Nucleoli contain RNA.

4. Examining Data Did each nucleus have only one nucleolus or several? What appeared to be the most common number of nucleoli?

Each nucleus will have at least one nucleolus, but most will have two or

three.

5. Describing Events Were you able to see any cells in the process of mitotic division? If so, what did the cells look like?

Students should be able to locate several cells in the process of mitotic

division. The nuclear material will appear to have condensed into short dark

chromosomes stained blue.

Conclusions

1. Drawing Conclusions Where is DNA located in both plant and animal cells?

DNA is found in the same location in both plant and animal cells; DNA is

found only in the chromosomes of the nucleus.

2. Drawing Conclusions Where is RNA located in both plant and animal cells?

RNA is found in the same locations in both plant and animal cells; RNA is

found in the nucleus, nucleoli, and cytoplasm.

Extension

Research and Communications Use library references to research other staining techniques.

Answer Key

Directed Reading

SECTION: MEIOSIS

1. Meiosis is a process that occurs during cell division that produces reproductive cells, such as gametes and spores. Meiosis involves two divisions of the nucleus, which halves the chromosome number of the cells.
2. Meiosis I involves the separation of pairs of homologous chromosomes. Meiosis II involves the separation of the two chromatids that make up each chromosome.
3. prophase I, metaphase I, anaphase I, telophase I, prophase II, metaphase II, anaphase II, and telophase II
4. Crossing-over is the exchange of reciprocal portions of DNA between two homologous chromosomes at the beginning of meiosis.
5. anaphase II
6. anaphase I
7. prophase I
8. prophase II
9. metaphase II
10. telophase II
11. telophase I
12. metaphase I
13. Crossing-over occurs when portions of a chromatid on one homologous chromosome are broken and exchanged with the corresponding chromatid portions of the other homologous chromosomes. Crossing-over takes place during prophase I of meiosis.
14. Independent assortment refers to the random distribution of homologous chromosomes during meiosis. Each of the 23 pairs of chromosomes separates independently during metaphase I and metaphase II. This means that one cell can produce many different gene combinations.
15. Spermatogenesis is the process by which sperm are produced in males. Oogenesis is the process by which eggs are produced in females.
16. Undifferentiated sperm cells are produced by meiosis. These cells become sperm by changing in form and developing tails.
17. In males, the cytoplasm divides equally following meiosis I and meiosis II to form four cells of the same size. In females, the cytoplasm divides unequally after meiosis I and meiosis II to produce one large egg and three smaller polar bodies.

SECTION: SEXUAL REPRODUCTION

1. In asexual reproduction, a single parent passes copies of all its genes to its offspring, and there is no fusion of gametes. In sexual reproduction, two parents each form gametes, which fuse to form offspring.
2. A clone is an offspring that is genetically identical to its parent. Asexual reproduction is the type of reproduction that produces clones.
3. Fission is the type of asexual reproduction that occurs in prokaryotes. Budding is a type of asexual reproduction in which new individuals split off from existing ones.
4. In budding, a new individual forms from another and may or may not break away from its parent. In fragmentation, a multicellular individual breaks into several pieces that may later develop into complete individuals by growing their missing parts.
5. Asexual reproduction allows organisms to produce many offspring in a short period of time, without using energy to produce gametes or find a mate. However, the DNA of these offspring varies little between individuals, and the organisms may not be able to adapt to a new environment if the environment changes.
6. Sexual reproduction provides a way of quickly making different combinations of genes among individuals. This genetic diversity may increase the ability of a species to adapt and succeed. But sexual reproduction requires energy to produce gametes and find a mate.

7. haploid life cycle (some fungi and algae, such as *Chlamydomonas*), diploid life cycle (most animals, including humans), alternation of generations (plants, such as roses)
8. The type of life cycle that a eukaryotic organism has depends on the type of cell that undergoes meiosis and on when meiosis occurs. Haploid cells occupy the major portion of the haploid life cycle. Diploid individuals occupy the major portion of the diploid life cycle. The gametophyte and the sporophyte take turns in the alternation of generations life cycle.
9. meiosis, gametes, fertilization, zygote, diploid individual

Active Reading

SECTION: MEIOSIS
1. TI
2. PI
3. AI
4. MI
5. PI
6. TI
7. AI
8. PI
9. AII
10. PII
11. TII
12. MII
13. TII
14. AII
15. TII
16. a

SECTION: SEXUAL REPRODUCTION
1. Reproduction, the process of producing offspring, can be asexual or sexual.
2. an organism that is genetically identical to its parent
3. binary fission
4. Because these offspring receive genetic material from both parents, they inherit traits from each.
5. In both processes, offspring are produced.
6. Because asexual reproduction involves a single parent, there is no fusion of haploid cells. Because sexual reproduction involves two parents, haploid cells are joined together.
7. c

Vocabulary Review

ACROSS
6. SPERMATOGENESIS
7. SEXUAL
9. OVUM
11. CROSSING OVER
12. SPORE
13. MEIOSIS
14. INDEPENDENT

DOWN
1. GAMETOPHYTE
2. FERTILIZATION
3. CLONE
4. ASEXUAL
5. LIFE
7. SPERM
8. SPOROPHYTE
10. OOGENESIS

Science Skills

SEQUENCING/ORGANIZING INFORMATION
1. g
2. b
3. c
4. f
5. h
6. d
7. a
8. e
9. g
10. a
11. d
12. f
13. c
14. h
15. e
16. b

Concept Mapping
1. meiosis II
2. haploid reproductive cells or gametes
3. crossing-over
4. homologous chromosomes
5. chromatids

Critical Thinking
1. g
2. f
3. b
4. e
5. a
6. h
7. c
8. d
9. b
10. a
11. c
12. d
13. b
14. d
15. c
16. a
17. d, c
18. h, g
19. b, f
20. a, e
21. d
22. c
23. b
24. d

Test Prep Pretest

1. genetic
2. four
3. budding, fission
4. mitosis, fusion
5. crossing, over
6. oogenesis
7. gametophyte
8. evolution
9. spores
10. genetic recombination
11. meiosis
12. prophase I
13. telophase I, cytokinesis
14. metaphase II
15. The formation of male and female gametes involves meiosis. In spermatogenesis, meiosis produces four sperm cells. In oogenesis, however, the cytoplasm divides unequally and meiosis produces three polar bodies that do not survive and one egg cell.
16. In fission, the parent divides into two or more individuals of the same size. In fragmentation, the parent's body breaks into several pieces that grow into complete individuals. In budding, new individuals split off from existing ones. The new individuals may be independent or attached to the parent.
17. In anaphase I, the two chromatids of each chromosome remain attached at their centromeres as the homologous pairs move to the poles of the cell. In anaphase II, the centromeres divide and the chromatids move to opposite poles of the cell. If the centromeres did not divide and split the paired chromosomes during anaphase II, the new cells would not be haploid.
18. In the haploid life cycle, haploid cells develop into haploid multicellular individuals that produce gametes through mitosis. A diploid zygote forms through fusion of opposite gametes, undergoes meiosis immediately, and creates new haploid cells. In the diploid life cycle, haploid gametes are produced through meiosis, and they create a diploid zygote through fertilization. The individual that results from the zygote remains diploid and occupies most of the diploid life cycle.
19. An advantage of sexual reproduction is that is provides a powerful means of quickly making different combinations of genes among individuals. This genetic diversity may increase the ability of a species to adapt and succeed. A disadvantage is that it requires energy to produce gametes and find a mate.
20. Crossing-over increases the numerical possibilities of genetic recombinations in individuals. Genetic diversity is what enables some members of a population to survive changing conditions. Increased genetic variation seems to increase the pace of evolution.

Quiz

SECTION: MEIOSIS

1. c
2. d
3. b
4. b
5. d
6. d
7. a
8. e
9. c
10. b

SECTION: SEXUAL REPRODUCTION

1. a
2. a
3. d
4. c
5. d
6. e
7. c
8. d
9. b
10. a

Chapter Test (General)

1. a
2. d
3. a
4. b
5. c
6. c
7. d
8. b
9. b
10. d
11. g
12. i
13. a
14. h
15. e
16. b
17. k
18. f
19. c
20. j

Chapter Test (Advanced)

1. b
2. d
3. c
4. a
5. d
6. a

7. Crossing-over results in the exchange of corresponding segments of DNA between homologous chromosomes. This results in genetic recombination.
8. In independent assortment in humans, each of the 23 pairs of chromosomes separates independently. Thus, about 8 million gametes with different gene combinations can be produced from one original cell. DNA is exchanged during crossing-over. This results in chromosomes in which the two chromatids no longer have identical genetic material.
9. Sexual reproduction creates genetic recombinations that may change the characteristics of the organisms. If there has been no change in environmental conditions that would require an adaptation by the organism, the new combinations of genes might harm the individual's ability to survive. Sexual reproduction also requires energy to find a mate and produce gametes.
10. Plants alternate between a diploid and a haploid phase. Growth of a zygote produces a multicellular sporophyte. Through meiosis, cells in the sporophyte produce haploid reproductive cells called spores. The spores grow into multicellular gametophytes. Gametophytes produce haploid gametes by mitosis, which fuse and develop into zygotes.
11. Spermatogenesis and oogenesis both produce gametes. Spermatogenesis occurs in males in the testes; oogenesis occurs in females in the ovaries. Spermatogenesis results in four haploid sperm cells. Oogenesis results in one haploid ovum and three polar bodies, which eventually die.
12. In asexual reproduction, a single parent passes copies of all of its genes to each of its offspring. In contrast, in sexual reproduction, two parents each form reproductive cells (gametes) that have one-half the number of chromosomes. These gametes join to form a diploid offspring.
13. m
14. h
15. d
16. i
17. l
18. a
19. k
20. f
21. j
22. c
23. b
24. g
25. e

TEACHER RESOURCE PAGE

Lesson Plan

Section: Meiosis

Pacing

Regular Schedule: with lab(s): 4 days   without lab(s): 3 days

Block Schedule: with lab(s): 2 days   without lab(s): 1 1/2 days

Objectives

1. Summarize the events that occur during meiosis.
2. Relate crossing-over, independent assortment, and random fertilization to genetic variation.
3. Compare spermatogenesis and oogenesis.

National Science Education Standards Covered

UNIFYING CONCEPTS AND PROCESSES

UCP1: Systems, order, and organization

UCP2: Evidence, models, and explanation

UCP3: Change, constancy, and measurement

UCP4: Evolution and equilibrium

UCP5: Form and function

SCIENCE AS INQUIRY

SI1: Abilities necessary to do scientific inquiry

SI2: Understandings about scientific inquiry

LIFE SCIENCE: THE CELL

LSCell1: Cells have particular structures that underlie their functions.

LSCell3: Cells store and use information to guide their functions.

LSCell4: Cell functions are regulated.

LSCell6: Cells can differentiate, and complex multicellular organisms are formed as a highly organized arrangement of differentiated cells.

LIFE SCIENCE: MOLECULAR BASIS OF HEREDITY

LSGene1: In all organisms, the instructions for specifying the characteristics of the organism are carried in DNA.

LSGene2: Most of the cells in a human contain two copies of each of 22 different chromosomes. In addition, there is a pair of chromosomes that determine sex.

Copyright © by Holt, Rinehart and Winston. All rights reserved.

Holt Biology   Meiosis and Sexual Reproduction

TEACHER RESOURCE PAGE

Lesson Plan *continued*

LSGene3: Changes in DNA (mutations) occur spontaneously at low rates.

LIFE SCIENCE: BIOLOGICAL EVOLUTION

LSEvol1: Species evolve over time.

LSEvol3: Natural selection and its evolutionary consequences provide a scientific explanation for the fossil record of ancient life forms as well as for the striking molecular similarities observed among the diverse species of living organisms.

KEY
SE = Student Edition   TE = Teacher Edition
CRF = Chapter Resource File

Block 1

CHAPTER OPENER *(45 minutes)*

- **Quick Review,** SE. Students answer questions covered in previous sections of the textbook as preparation for the chapter content. **(GENERAL)**
- **Reading Activity,** SE. Students study the first two pages of the chapter and then answer questions on their own paper. **(GENERAL)**
- **Using the Figure,** TE. Students answer questions about the chapter opener photograph. **(GENERAL)**
- **Opening Activity,** Multiplying Chromosomes, TE. Students calculate the chromosome number in three different organisms over eight generations, assuming that the chromosome number doubles each generation. **(GENERAL)**

Block 2

FOCUS *(5 minutes)*

- **Bellringer Transparency.** Use this transparency as students enter the classroom and find their seats. **(GENERAL)**

MOTIVATE *(10 minutes)*

- **Demonstration,** TE. Use this simple demonstration on the board or using an overhead transparency to introduce the concept of meiosis. **(GENERAL)**

TEACH *(30 minutes)*

- **Teaching Transparency, Section Outline.** Use this transparency to give students a framework for the information in this section. **(GENERAL)**
- **Directed Reading Worksheet, Meiosis, CRF.** Students complete the exercises in this worksheet to help them understand the material as they read the section. **(BASIC)**
- **Teaching Transparency, Stages of Meiosis.** Use this transparency to walk students through the steps of meiosis. **(GENERAL)**

TEACHER RESOURCE PAGE

Lesson Plan continued

- **Group Activity,** Modeling Meiosis, TE. Students use beads and pipe cleaners to model meiosis. (GENERAL)

HOMEWORK

- **Active Reading Worksheet, Meiosis, CRF.** Students read a passage related to the section topic and answer questions. (GENERAL)

Block 3

TEACH *(30 minutes)*

- **Quick Lab,** Modeling Crossing-Over, SE. Students use paper models to compare the number of different types of chromatids before crossing-over with the number after crossing-over. (GENERAL)
- **Teaching Transparency, Meiosis in Male and Female Animals.** Use this transparency to compare meiosis in males and females. (GENERAL)
- **Teaching Tip,** Distinguishing Between Mitosis and Meiosis, TE. Students make a graphic organizer to summarize the difference in chromosomal number between mitosis and meiosis. (GENERAL)

CLOSE *(15 minutes)*

- **Reteaching,** TE. Students use prepared paper models of homologous chromosome pairs to model meiosis and crossing-over. Ask students why sexual reproduction provides more diversity in organisms. (BASIC)
- **Quiz,** TE. Students answer questions that review the section material. (GENERAL)

HOMEWORK

- **Section Review,** SE. Assign questions 1–6 for review, homework, or quiz. (GENERAL)
- **Alternative Assessment,** TE. Students write a story from the viewpoint of a single chromosome. (GENERAL)
- **Quiz, CRF.** This quiz consists of ten multiple choice and matching questions that review the section's main concepts. (BASIC) Also in Spanish.

Optional Block

LAB *(45 minutes)*

- **Skills Practice Lab, Laboratory Techniques: Staining DNA and RNA, CRF.** Students use methyl green-pyronin Y stain for detecting both DNA and RNA. (GENERAL)

TEACHER RESOURCE PAGE

Lesson Plan *continued*

Other Resource Options

- **Internet Connect.** Students can research Internet sources about Genetic Variation with SciLinks Code HX4093.
- **Internet Connect.** Students can research Internet sources about Oogenesis with SciLinks Code HX4192.
- **Internet Connect.** Students can research Internet sources about Meiosis with SciLinks Code HX4120.
- **go.hrw.com.** For worksheets, videos, and other teaching aids related to this chapter, visit the HRW Web site and type in the keyword HX4 MEI.
- **CNN Science in the News, Video Segment 4 Organ Cloning.** This video segment is accompanied by a **Critical Thinking Worksheet**.
- **CNN Student News.** Find the latest news, lesson plans, and activities related to important scientific events at **cnnstudentnews.com**.
- **Biology Interactive Tutor CD-ROM,** Unit 4 Cell Reproduction. Students watch animations and other visuals as the tutor explains meiosis. Students assess their learning with interactive activities.

Lesson Plan

Section: Sexual Reproduction

Pacing

Regular Schedule: with lab(s): 4 days   without lab(s): 2 days
Block Schedule: with lab(s): 2 days   without lab(s): 1 day

Objectives

1. Differentiate between asexual and sexual reproduction.
2. Identify three types of asexual reproduction.
3. Evaluate the relationship and evolutionary advantages and disadvantages of asexual and sexual reproduction.
4. Differentiate between the three major sexual life cycles found in eukaryotes.

National Science Education Standards Covered

UNIFYING CONCEPTS AND PROCESSES

UCP1: Systems, order, and organization

UCP5: Form and function

SCIENCE AS INQUIRY

SI1: Abilities necessary to do scientific inquiry

SI2: Understandings about scientific inquiry

LIFE SCIENCE: THE CELL

LSCell1: Cells have particular structures that underlie their functions.

LSCell3: Cells store and use information to guide their functions.

LSCell4: Cell functions are regulated.

LSCell6: Cells can differentiate and form complete multicellular organisms.

LIFE SCIENCE: MOLECULAR BASIS OF HEREDITY

LSGene1: In all organisms, the instructions for specifying the characteristics of the organisms are carried in DNA.

LSGene2: Most of the cells in a human contain two copies of each of 22 different chromosomes. In addition, there is a pair of chromosomes that determines sex.

LSGene3: Changes in DNA (mutations) occur spontaneously at low rates.

TEACHER RESOURCE PAGE

Lesson Plan *continued*

> **KEY**
> SE = Student Edition   TE = Teacher Edition
> CRF = Chapter Resource File

Block 4

FOCUS *(5 minutes)*

- **Bellringer Transparency.** Use this transparency as students enter the classroom and find their seats. **(GENERAL)**

MOTIVATE *(10 minutes)*

- **Identifying Preconceptions**, TE. Students discuss the difference between asexual and sexual reproduction. **(GENERAL)**

TEACH *(30 minutes)*

- **Teaching Transparency, Section Outline.** Use this transparency to give students a framework for the information in this section. **(GENERAL)**
- **Quick Lab,** Observing Reproduction in Yeast, SE. Students examine a yeast culture to observe one of the types of reproduction that yeast undergoes. **(GENERAL)**
- **Datasheets for In-Text Labs,** Observing Reproduction in Yeast, CRF.
- **Activity,** Asexual Reproduction in Bacteria, TE. Students calculate the number of bacteria produced from a single bacterium after each hour for six hours. **(GENERAL)**

HOMEWORK

- **Directed Reading Worksheet,** Sexual Reproduction, CRF. Students complete the exercises in this worksheet to help them understand the material as they read the section. **(BASIC)**

Block 5

TEACH *(25 minutes)*

- **Active Reading Worksheet,** Sexual Reproduction, CRF. Students read a passage related to the section topic and answer questions. **(GENERAL)**
- **Teaching Transparency, Haploid Life Cycle.** Use this transparency to walk students through the main stages in a haploid life cycle. **(GENERAL)**
- **Teaching Transparency, Diploid Life Cycle.** Use this transparency to walk students through the main stages in a diploid life cycle. **(GENERAL)**
- **Teaching Transparency, Alternation of Generations.** Use this transparency to show students that alternation of generations alternates between diploid and haploid phases. **(GENERAL)**

Copyright © by Holt, Rinehart and Winston. All rights reserved.

Holt Biology   Meiosis and Sexual Reproduction

TEACHER RESOURCE PAGE

Lesson Plan *continued*

CLOSE *(20 minutes)*

- **Alternative Assessment,** TE. Students make a chart showing the advantages and disadvantages of asexual and sexual reproduction. **(GENERAL)**
- **Reteaching,** TE. Students discuss answers to these questions as a review of the section material. **(BASIC)**
- **Quiz,** TE. Students answer questions that review the section material. **(GENERAL)**

HOMEWORK

- **Section Review,** SE. Assign questions 1–6 for review, homework, or quiz. **(GENERAL)**
- **Science Skills Worksheet, CRF.** Students sequence the stages of meiosis and answer questions about the process. **(GENERAL)**
- **Quiz, CRF.** This quiz consists of ten multiple choice and matching questions that review the section's main concepts. **(BASIC) Also in Spanish.**
- **Modified Worksheet, One-Stop Planner.** This worksheet has been specially modified to reach struggling students. **(BASIC)**
- **Critical Thinking Worksheet, CRF.** Students answer analogy-based questions that review the section's main concepts and vocabulary. **(ADVANCED)**

Optional Block

LAB *(45 minutes)*

- **Exploration Lab,** Modeling Meiosis, SE. Students model meiosis and then use their model to answer questions about the process. **(GENERAL)**
- **Datasheets for In-Text Labs,** Modeling Meiosis, CRF.

Other Resource Options

- **Exploration Lab, Karyotyping, CRF.** Students complete and analyze a karyotype of fetal cells to determine the sex of the fetus and whether the fetus is normal or has Down syndrome. **(GENERAL)**
- **Supplemental Reading, The Lives of a Cell, One-Stop Planner.** Students read the book and answer questions. **(ADVANCED)**
- **Exploring Further,** Cloning by Parthenogenesis, TE. Use these questions to lead a discussion about the article in the SE. **(ADVANCED)**
- **go.hrw.com.** For worksheets, videos, and other teaching aids related to this chapter, visit the HRW Web site and type in the keyword HX4 MEI.
- **CNN Student News.** Find the latest news, lesson plans, and activities related to important scientific events at **cnnstudentnews.com**.
- **CNN Science in the News, Video Segment 4 Organ Cloning.** This video segment is accompanied by a **Critical Thinking Worksheet**.

TEACHER RESOURCE PAGE

Lesson Plan

End-of-Chapter Review and Assessment

Pacing
Regular Schedule: 2 days
Block Schedule: 1 day

KEY
SE = Student Edition   TE = Teacher Edition
CRF = Chapter Resource File

Block 6
REVIEW *(45 minutes)*

_ **Study Zone,** SE. Use the Study Zone to review the Key Concepts and Key Terms of the chapter and prepare students for the Performance Zone questions. **(GENERAL)**

_ **Performance Zone,** SE. Assign questions to review the material for this chapter. Use the assignment guide to customize review for sections covered. **(GENERAL)**

_ **Teaching Transparency, Concept Mapping.** Use this transparency to review the concept map for this chapter. **(GENERAL)**

Block 7
ASSESSMENT *(45 minutes)*

_ **Chapter Test, Meiosis and Sexual Reproduction, CRF.** This test contains 20 multiple choice and matching questions keyed to the chapter's objectives. **(GENERAL) Also in Spanish.**

_ **Chapter Test, Meiosis and Sexual Reproduction, CRF.** This test contains 25 questions of various formats, each keyed to the chapter's objectives. **(ADVANCED)**

_ **Modified Chapter Test, One-Stop Planner.** This test has been specially modified to reach struggling students. **(BASIC)**

Other Resource Options

_ **Vocabulary Review Worksheet, CRF.** Use this worksheet to review the chapter vocabulary. **(GENERAL) Also in Spanish.**

_ **Test Prep Pretest, CRF.** Use this pretest to review the main content of the chapter. Each question is keyed to a section objective. **(GENERAL) Also in Spanish.**

_ **Test Item Listing for ExamView® Test Generator, CRF.** Use the Test Item Listing to identify questions to use in a customized homework, quiz, or test.

_ **ExamView® Test Generator, One-Stop Planner.** Create a customized homework, quiz, or test using the HRW Test Generator program.

TEST ITEM LISTING
Meiosis and Sexual Reproduction

TRUE/FALSE

1. ____ While paired together during the second division of meiosis, two chromosomes may exchange segments of DNA.
   Answer: False   Difficulty: I   Section: 1   Objective: 1

2. ____ Meiosis produces four nuclei that have different chromosome numbers from the parent cell.
   Answer: True   Difficulty: I   Section: 1   Objective: 1

3. ____ Crossing-over is the exchange of corresponding portions of chromatids between homologous chromosomes.
   Answer: True   Difficulty: I   Section: 1   Objective: 2

4. ____ At the conclusion of crossing-over, genetic recombination has occurred.
   Answer: True   Difficulty: I   Section: 1   Objective: 2

5. ____ Independent assortment occurs when each pair of chromosomes segregates (separates) independently.
   Answer: True   Difficulty: I   Section: 1   Objective: 2

6. ____ Random fertilization refers to the fact that gametes are produced independently.
   Answer: False   Difficulty: I   Section: 1   Objective: 2

7. ____ The process by which sperm are produced in male animals is called spermatogenesis.
   Answer: True   Difficulty: I   Section: 1   Objective: 3

8. ____ Oogenesis occurs in the female reproductive organs.
   Answer: True   Difficulty: I   Section: 1   Objective: 3

9. ____ Gametogenesis occurs only in males.
   Answer: False   Difficulty: I   Section: 1   Objective: 3

10. ____ Meiosis in female animals results in the same number of ova as sperm that were produced by meiosis in the male.
    Answer: False   Difficulty: I   Section: 1   Objective: 3

11. ____ The two cells produced during the first cytokinesis in female animals are approximately equal in size and contain the same amount of cytoplasm.
    Answer: False   Difficulty: I   Section: 1   Objective: 3

12. ____ Some organisms look exactly like their parents.
    Answer: True   Difficulty: I   Section: 2   Objective: 1

13. ____ In asexual reproduction, two parents each pass copies of all of their cells to their offspring.
    Answer: False   Difficulty: I   Section: 2   Objective: 1

14. ____ In sexual reproduction, two parents each form haploid cells, which join to form offspring.
    Answer: True   Difficulty: I   Section: 2   Objective: 1

15. ____ Some eukaryotes reproduce asexually, and some reproduce sexually.
    Answer: True   Difficulty: I   Section: 2   Objective: 1

16. ____ Amoebas reproduce by fission.
    Answer: True   Difficulty: I   Section: 2   Objective: 2

Copyright © by Holt, Rinehart and Winston. All rights reserved.

TEST ITEM LISTING, continued

17. ____ In budding, new individuals develop from fragments of the original individual.
    Answer: False   Difficulty: I   Section: 2   Objective: 2

18. ____ Asexual reproduction provides for genetic diversity, the raw material for evolution.
    Answer: False   Difficulty: I   Section: 2   Objective: 3

19. ____ The pairing of homologous chromosomes during meiosis may have originally been a way to repair damaged DNA.
    Answer: True   Difficulty: I   Section: 2   Objective: 3

20. ____ Genetic diversity is the raw material for evolution.
    Answer: True   Difficulty: I   Section: 2   Objective: 3

21. ____ In most animals, including humans, meiosis produces sperm and egg cells.
    Answer: True   Difficulty: I   Section: 2   Objective: 4

22. ____ Alternation of generations is the simplest of the sexual life cycles.
    Answer: False   Difficulty: I   Section: 2   Objective: 4

23. ____ During the haploid life cycle, the zygote is the only diploid cell.
    Answer: True   Difficulty: I   Section: 2   Objective: 4

24. ____ During the diploid life cycle, all of the cells are diploid.
    Answer: False   Difficulty: I   Section: 2   Objective: 4

25. ____ Plants, algae, and some protists have a life cycle that regularly alternates between a haploid phase and a diploid phase.
    Answer: True   Difficulty: I   Section: 2   Objective: 4

26. ____ In plants, the diploid phase in the life cycle that produces spores is called a gametophyte.
    Answer: False   Difficulty: I   Section: 2   Objective: 4

27. ____ In the life cycle of a plant, the gametophyte is the haploid phase that produces gametes by mitosis.
    Answer: True   Difficulty: I   Section: 2   Objective: 4

28. ____ Unlike a gamete, a spore gives rise to a multicellular individual without joining with another cell.
    Answer: True   Difficulty: I   Section: 2   Objective: 4

29. ____ Roses are examples of plants that have a life cycle called alternation of generations.
    Answer: True   Difficulty: I   Section: 2   Objective: 4

30. ____ Moss plants have haploid life cycles.
    Answer: False   Difficulty: I   Section: 2   Objective: 4

TEST ITEM LISTING, continued

MULTIPLE CHOICE

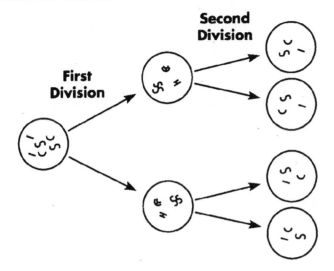

31. Refer to the illustration above. The process shown is
    a. mitosis.
    b. chromosomal mutation.
    c. meiosis.
    d. dominance.
    Answer: C   Difficulty: II   Section: 1   Objective: 1

32. Separation of homologues occurs during
    a. mitosis.
    b. meiosis I.
    c. meiosis II.
    d. fertilization.
    Answer: B   Difficulty: I   Section: 1   Objective: 1

33. The difference between anaphase of mitosis and anaphase I of meiosis is that
    a. the chromosomes line up at the equator in anaphase I.
    b. centromeres do not exist in anaphase I.
    c. chromatids do not separate at the centromere in anaphase I.
    d. crossing-over occurs only in anaphase of mitosis.
    Answer: C   Difficulty: II   Section: 1   Objective: 1

34. When crossing-over takes place, chromosomes
    a. mutate in the first division.
    b. produce new genes.
    c. decrease in number.
    d. exchange corresponding segments of DNA.
    Answer: D   Difficulty: I   Section: 1   Objective: 1

Using the information you have learned about cell reproduction, infer answers to the questions below about a cell with a diploid number of 4 chromosomes. Select from among the diagrams below, labeled A, B, C, D, and E, to answer the questions.

Copyright © by Holt, Rinehart and Winston. All rights reserved.

Holt Biology   T 3   Meiosis and Sexual Reproduction

TEST ITEM LISTING, continued

35. Which of the diagrams above depicts a cell at the beginning of mitosis?
    a. B
    b. C
    c. D
    d. E

    Answer: C   Difficulty: II   Section: 1   Objective: 1

36. Which of the diagrams above depicts a cell at the end of meiosis I?
    a. B
    b. C
    c. D
    d. E

    Answer: D   Difficulty: III   Section: 1   Objective: 1

37. Which of the diagrams above depicts a cell at the end of meiosis II?
    a. A
    b. B
    c. C
    d. D

    Answer: A   Difficulty: II   Section: 1   Objective: 1

38. Which of the diagrams above depicts a cell at the end of mitosis?
    a. A
    b. B
    c. C
    d. D

    Answer: B   Difficulty: II   Section: 1   Objective: 1

39. The exchange of segments of DNA between the members of a pair of chromosomes
    a. ensures that variations within a species never occur.
    b. acts as a source of variations within a species.
    c. always produces genetic disorders.
    d. is called crossing.

    Answer: B   Difficulty: I   Section: 1   Objective: 2

40. Crossing-over occurs
    a. during prophase II.
    b. during fertilization.
    c. during prophase I.
    d. at the centromere.

    Answer: C   Difficulty: I   Section: 1   Objective: 2

41. Which of the following does *not* provide new genetic combinations?
    a. random fertilization
    b. cytokinesis
    c. independent assortment
    d. crossing-over

    Answer: B   Difficulty: I   Section: 1   Objective: 2

42. During cytokinesis in the female, what divides unequally?
    a. the sperm cell
    b. the ovary
    c. the cytoplasm
    d. None of the above

    Answer: C   Difficulty: I   Section: 1   Objective: 3

43. The more common name for an ovum is a(n)
    a. egg.
    b. hormone.
    c. nutrient.
    d. polar body.

    Answer: A   Difficulty: I   Section: 1   Objective: 3

44. The process of producing offspring is called reproduction and can be
    a. eukaryotic or prokaryotic.
    b. asexual or sexual.
    c. cardiovascular or respiratory.
    d. None of the above

    Answer: B   Difficulty: I   Section: 2   Objective: 1

45. Which of the following is *not* a type of asexual reproduction?
    a. budding
    b. fragmentation
    c. fission
    d. fertilization

    Answer: D   Difficulty: I   Section: 2   Objective: 2

TEST ITEM LISTING, continued

46. Types of asexual reproduction include
    a. budding.
    b. fragmentation.
    c. fission.
    d. All of the above
    Answer: D   Difficulty: I   Section: 2   Objective: 2

47. Hydras reproduce by
    a. budding.
    b. fragmentation.
    c. fission.
    d. None of the above
    Answer: B   Difficulty: I   Section: 2   Objective: 2

48. Budding is an example of
    a. endosymbiosis.
    b. asexual reproduction.
    c. meiosis.
    d. sexual reproduction.
    Answer: B   Difficulty: I   Section: 2   Objective: 2

49. The simplest and most primitive method of reproduction is
    a. sexual.
    b. diploid.
    c. haploid.
    d. None of the above
    Answer: D   Difficulty: I   Section: 2   Objective: 3

50. Which of the following is *not* a type of sexual life cycle?
    a. alternation of generations
    b. haploid
    c. diploid
    d. cellular
    Answer: D   Difficulty: I   Section: 2   Objective: 4

51. To create new haploid cells during the haploid life cycle, the zygote undergoes
    a. mitosis.
    b. fertilization.
    c. fusion.
    d. meiosis.
    Answer: D   Difficulty: I   Section: 2   Objective: 4

52. In alternation of generations, a diploid spore-forming cell gives rise to four
    a. zygotes.
    b. sperm cells.
    c. haploid spores.
    d. diploid spores.
    Answer: C   Difficulty: I   Section: 2   Objective: 4

53. During alternation of generations, cells reproduce by
    a. meiosis.
    b. mitosis.
    c. both meiosis and mitosis.
    d. None of the above
    Answer: C   Difficulty: I   Section: 2   Objective: 4

COMPLETION

54. The stage of meiosis during which homologues line up along the equator of the cell is called _____ _____.
    Answer: metaphase I   Difficulty: II   Section: 1   Objective: 1

55. Fertilization of the haploid sperm and egg results in the restoration of the _____ number of chromosomes in the zygote.
    Answer: diploid   Difficulty: II   Section: 1   Objective: 1

56. After a new nuclear membrane forms during telophase of meiosis, the _____ divides, resulting in two cells.
    Answer: cytoplasm   Difficulty: II   Section: 1   Objective: 1

57. The cells resulting from meiosis in either males or females are called _____.
    Answer: gametes   Difficulty: II   Section: 1   Objective: 1

TEST ITEM LISTING, continued

58. The process called _____ guarantees that the number of chromosomes in gametes is half the number of chromosomes in body cells.
    Answer: meiosis   Difficulty: II   Section: 1   Objective: 1

59. A reciprocal exchange of corresponding segments of DNA is called _____-_____.
    Answer: crossing-over   Difficulty: II   Section: 1   Objective: 1

60. The four haploid cells formed in the male at the end of meiosis II develop a tail and are called _____.
    Answer: sperm   Difficulty: II   Section: 1   Objective: 3

61. An individual produced by asexual reproduction that is genetically identical to its parent is called a(n) _____.
    Answer: clone   Difficulty: II   Section: 2   Objective: 1

62. The separation of a parent into two or more individuals of about equal size is called _____.
    Answer: fission   Difficulty: II   Section: 2   Objective: 2

63. The process in which sperm and egg cells join is called _____.
    Answer: fertilization   Difficulty: II   Section: 2   Objective: 4

64. A spore is a haploid reproductive cell produced by _____.
    Answer: meiosis   Difficulty: II   Section: 2   Objective: 4

65. The entire span in the life of an organism from one generation to the next is called a(n) _____ _____.
    Answer: life cycle   Difficulty: II   Section: 2   Objective: 4

66. The diploid phase in the life cycle of plants is called the _____.
    Answer: sporophyte   Difficulty: II   Section: 2   Objective: 4

ESSAY

67. What would happen if the chromosome number were not reduced before sexual reproduction?
    Answer:
    The number of chromosomes in the offspring would be double the number in the parents. The number and characteristics of chromosomes in cells determine the traits of the organism. The organism would almost certainly not survive the doubling of its chromosomes, and even if it did survive and reproduce, then the number of chromosomes would become unmanageably large after only a few generations.
    Difficulty: III   Section: 1   Objective: 1

68. Compare the features of mitotic metaphase, meiotic metaphase I, and meiotic metaphase II.
    Answer:
    During metaphase of mitosis, the diploid number of chromosomes of the cell line up single file across the equator of the cell. Meiotic metaphase I is characterized by the homologous chromosomes lining up as pairs (double file) along the equator. Metaphase II of meiosis appears similar to mitotic metaphase, except that the number of chromosomes is the haploid number rather than the diploid number. These chromosomes line up single file across the cell equator.
    Difficulty: III   Section: 1   Objective: 1

TEST ITEM LISTING, continued

69. Identify three ways in which genetic recombination results during meiosis.
    Answer:
    Genetic recombination results when crossing-over occurs between homologous chromosomes, when homologous pairs separate independently in meiosis I, when sister chromatids separate independently in meiosis II, and when the zygote that forms a new individual is created by the random joining of two gametes.
    Difficulty: III   Section: 1   Objective: 2

70. Explain why crossing-over is an important source of genetic variation.
    Answer:
    Crossing-over occurs when two homologous chromosomes exchange reciprocal segments of DNA during prophase I of meiosis. This results in chromosomes in which the two chromatids no longer have identical genetic material. When meiosis is completed, the resulting gametes carry new combinations of genes.
    Difficulty: III   Section: 1   Objective: 2

71. What are the two things that might happen to a bud?
    Answer:
    It may break from the parent and become an independent organism, or it may remain attached to the parent and eventually give rise to many individuals.
    Difficulty: III   Section: 2   Objective: 2

72. What are at least two advantages of asexual reproduction?
    Answer:
    1. It is less complex than sexual reproduction.
    2. In a stable environment, it allows organisms to produce many offspring in a short period of time.
    3. Organisms do not need to use energy to produce gametes or to find a mate.
    Difficulty: II   Section: 2   Objective: 3